This book is a gift for all early years' teachers; it is a gift they have been waiting for, as they ponder how to make a positive contribution to the troubled world the children are growing up in. Blom offers a way of thinking-in-being, where teachers, and the children they teach, are not outside nature, or opposite to nature. The children are not growing up *in* that world, but *as* that world; indeed, we are, as humans, not superior to nature, or separate from it, or above it; we are, all of us, nature.

– Professor Bronwyn Davies,
Professorial Fellow, University of
Melbourne and Emeritus Professor,
Western Sydney University

Blom begins this book positing that "there is a desperate and urgent need for monumental changes to education systems" and I think that is correct. So where better to start than from the beginning, accompanying young children and their human and more-than-human educators as they re-turn themselves into earthy relationality. A must-read for those teaching, thinking, and seeking in the direction of monumental change.

– Professor Sean Blenkinsop, *Faculty*
of Education, Simon Fraser University
and Co-Director, Imaginative Education
Research Group (IERG)

In this fascinating and challenging book, Blom takes cues from posthumanism, postqualitative inquiry, and non-representational theory in advancing a novel approach to educational research called transqualitative inquiry. But it doesn't stop there! Blom then takes us on a 'diffractive journey" through a series of data entanglements that demonstrate how transqualitative inquiry can be put to work. The result is a "deep dive" into the intricacies of everyday classroom practices and happenings which privileges the voices of the teachers and the nonhuman, and thus the "response-ability" of teachers to their students and the planet. As an important transdisciplinary text, this book is highly recommended for postgraduate research students in education but also researchers in cognate disciplines like cultural geography, cultural studies, anthropology, environmental psychology, and even nursing.

– Associate Professor Candice P. Boyd,
School of Geography, Earth and
Atmospheric Sciences, The University of
Melbourne, Australia

RETURNING LEARNING

Returning Learning explores early school years teachers' perceptions of nature and how this informs their pedagogy through a posthuman theoretical framework. The theoretical framework is purposefully designed to disrupt dichotomies and reject abuse to marginalised others. In doing so, this book offers a reconceptualisation of learning in environmental education, and education more broadly.

The posthuman theoretical framework is a transdisciplinary offering informed by material-discursive practices, affective atmospheres, and childhoodnature. The theoretical framework and transqualitative methodology support diffractive ethnographic methods where data are generated through an iterative and entangled data collection and data analysis process. This process is presented as a series of "diffractive data entanglements" that explore teachers' perceptions of nature, their pedagogical practices, and the implications of these data through a posthuman framing. These non-conventional approaches to undertaking research are the foundation for this book that listens to teacher's voices by conducting *research with* teachers rather than *to* teachers.

Through a deep exploration into the intricacies of everyday classroom practices and happenings, this book privileges the voices of the teachers and the nonhuman, thus the response-ability of teachers to their students and the planet, is re-turned. It will be of interest to researchers who are interested in creative and innovative theories and methodologies as well as those studying environmental education and other pedagogical studies as part of their courses.

Simone M. Blom, Southern Cross University, Australia.

Postqualitative, New Materialist and Critical Posthumanist Research

Editor in Chief:
Karin Murris
(Universities of Oulu, Finland, and Cape Town, South Africa)

Editors:
Vivienne Bozalek (University of the Western Cape and Rhodes University, South Africa)
Asilia Franklin-Phipps (State University of New York at New Paltz, USA)
Simone Fullagar (Griffith University, Australia)
Candace R. Kuby (University of Missouri, USA)
Karen Malone (Swinburne University of Technology, Australia)
Carol A. Taylor (University of Bath, United Kingdom)
Weili Zhao (Hangzhou Normal University, China)

This cutting-edge series is designed to assist established researchers, academics, postgraduate/graduate students, and their supervisors across higher education faculties and departments to incorporate novel, postqualitative, new materialist, and critical posthumanist approaches in their research projects and their academic writing. In addition to these substantive foci, books within the series are inter-, multi-, or transdisciplinary and are in dialogue with perspectives such as Black feminisms and Indigenous knowledges, decolonial, African, Eastern, and young children's philosophies. Although the series' primary aim is accessibility, its scope makes it attractive to established academics already working with postqualitative approaches.

This series is unique in providing short, user-friendly, affordable books that support postgraduate students and academics across disciplines and faculties in higher education. The series is supported by its own website with videos, images and other forms of 3D transmodal expression of ideas – provocations for research courses.

More resources for the books in the series are available on the series website, www.postqualitativeresearch.com.

If you.are interested in submitting a proposal for the series, please write to the Chief Editor, Professor Karin Murris: karin.murris@oulu.fi; karin.murris@uct.ac.za.

Other volumes in this series include:

For a full list of titles in this series, please visit: www.routledge.com/Postqualitative-New-Materialist-and-Critical-Posthumanist-Research/book-series/PNMR

RETURNING LEARNING

A Diffractive, Posthuman Exploration
of Nature Perceptions and Pedagogies
with Early School Years' Teachers

Simone M. Blom

Routledge
Taylor & Francis Group

LONDON AND NEW YORK

Designed cover image: Lauren Herman

First published 2025
by Routledge
4 Park Square, Milton Park, Abingdon, Oxon OX14 4RN

and by Routledge
605 Third Avenue, New York, NY 10158

Routledge is an imprint of the Taylor & Francis Group, an informa business

British Library Cataloguing-in-Publication Data
A catalogue record for this book is available from the British Library

Library of Congress Cataloging-in-Publication Data
Names: Blom, Simone M., author.
Title: Returning learning : a diffractive, posthuman exploration of nature
perceptions and pedagogies with early school years' teachers / Simone M.
Blom.
Description: Abingdon, Oxon ; New York : Routledge, 2025. | Series:
Postqualitative, new materialist and critical posthumanist research
series | Includes bibliographical references and index.
Identifiers: LCCN 2024035042 (print) | LCCN 2024035043 (ebook) | ISBN
9781032703411 (hardback) | ISBN 9781032691480 (paperback) | ISBN
9781032703473 (ebook)
Subjects: LCSH: Environmental education--Study and teaching (Early
childhood) | Posthumanism--Research. | Philosophy of nature. | Children
and the environment. | Culturally relevant pedagogy.
Classification: LCC GE70 .B58 2025 (print) | LCC GE70 (ebook) | DDC
372.35/7--dc23/eng/20240902
LC record available at https://lccn.loc.gov/2024035042
LC ebook record available at https://lccn.loc.gov/2024035043

ISBN: 9781032703411 (hbk)
ISBN: 9781032691480 (pbk)
ISBN: 9781032703473 (ebk)

DOI: 10.4324/9781032703473

Typeset in Optima
by KnowledgeWorks Global Ltd.

To all those that are on the journey,
may this work be of use.

CONTENTS

SERIES EDITOR FOREWORD

It is a delight to welcome the new book, *Returning Learning: A Diffractive, Posthuman Exploration of Nature Perceptions and Pedagogies with Early School Year's Teachers*, to the Postqualitative, New Materialist and Critical Posthumanist Research Series. The book is a persuasive mix of Simone Blom's personal and professional perspectives on environmental education. Diffracting complex philosophical, educational and political arguments through one another makes a compelling read.

The book starts from the premise that planetary catastrophes are real and have created a culture of fear, leaving people feeling disempowered and paralysed in thinking they are unable make a difference. Justifiably critical of the Anthropocene as a term "drenched in colonised thinking", she speculates about what an educational system could look and feel like that prioritises learning with or as nature, and not (as is currently the case), learning in or about nature "out there". This would urgently require an ontological shift in theory practice that doesn't position nature as separate from humans and equates nature with "wild" and "animal", as often is the case when researching with children. The neologism childhoodnature disrupts the nature/culture binary and makes it possible (also for children themselves) to explore human/nonhuman relationalities. The concept childhoodnature (posthuman child) reconfigures nature as teacher and child as nature.

The book starts by taking us on a systematically laid out journey. First, Simone Blom meanders competently through a wealth of relevant environmental education and posthumanist literature, thereby purposefully educating the reader in the necessity of disrupting power-producing binaries such as culture/nature, human/nonhuman, adult/child, parent/child, and theory/practice. This review of the literature is itself diffractive; she re-turns to her own learning about (environmental) learning by present/ing (threading) past literature to re-think future educational practices. The apparatuses of her proposed Returning Learning framework are the entangled concepts childhoodnature, material-discursive practices and affective atmospheres. Intra-actively, these concepts perform the de(con)struction of binaries, question human exceptionalism and tune into human-nonhuman entanglements. The Returning Learning framework is put to work in subsequent chapters by diffracting through the teaching of an everyday lesson by a First Nation classroom teacher. This convincingly illustrates Blom's proposal for a reconfigured environmental education that is transdisciplinary and moreover shows how research can be performed differently. Readers are even invited to join the intra-active analysis!

Blom's admiration, care and respect for teachers stands out. Teachers' own perceptions of "nature" and how these inform their pedagogical practices are at the heart of this well-written book. Perceptions are theorised as a broad term to include conceptualisations, attitudes, values, worldviews, and other socio-cultural dimensions of people's material-discursive beliefs. A teacher herself, Blom acknowledges how teachers are bound by rigid curricula and (too) many administrative duties. Her interest in teachers' own perceptions is fuelled by real curiosity in the "how" and "why" of their practices and is grounded in a genuine openness to find solutions that really work. This pragmatic dimension is one important reason why this book fits so well within the series. Teachers' perspectives tend to be undervalued – "silenced for too long" – and it is critical (also for their viability) that educational solutions include teachers' voice and professionalism. I was particularly struck by her acknowledgement that the Australian Early Years Learning Framework (EYLF) does consider human/nonhuman relations, but at the same time assumes a human-centric ontology. Blom's subtle and profound understanding of posthumanism(s) helps her put forward a radical proposal for different educational futures which is realistic because it is grounded in teachers' own beliefs and practices. The study foregrounds those ideas that are about "unsilencing teachers' own knowing along with that of First Nations' knowledges". Although the field work was carried out in Australia and the curricula referred to are Australian, the main arguments are relevant for education research elsewhere. One of the compelling questions for us to consider is: How does, or how can, environmental education be reconfigured when teachers know (they are) nature?

In terms of research methods, Blom invites the reader to consider transqualitative inquiry as a diffractive ethnographic research methodology. Transqualitative inquiry challenges qualitative research methods by moving away from describing (child) humans only (and the things they do) but should not be equated with "post-qualitative inquiry" as defined by Elizabeth St Pierre.

St Pierre prefers "inquiry", because for her the post-qualitative turn is against methods. Post-qualitative inquiry is a doing of post-structuralist theories and based on a flat ontology (a philosophy of immanence). Unlike St Pierre who finds methods "boring", Blom embraces methods. She proposes that, as a process, posthumanism is always re-configuring, re-turning, and re-inventing itself and focused on Baradian "phenomena" (not bounded objects or subjects). In practice, this means that the processes of data collection and analysis can't be separated out and are always transitional and entangled ontologically. Blom's reasons for disagreeing with St Pierre are unusual in a field that is divided about the use of methods and the controversial notion of a flat ontology. This makes the book even more interesting for readers of this series.

Originally a doctorate study, *Returning Learning* makes a unique contribution to knowledge in environmental education and posthumanist research. Altogether, another welcome addition to the series.

Karin Murris, Chief Editor of the Series

FOREWORD

This book is a gift for all early years' teachers; it is a gift they have been waiting for, as they ponder how to make a positive contribution to the troubled world the children are growing up in. Blom offers a way of thinking-in-being, where teachers, and the children they teach, are not outside nature, or opposite to nature. The children are not growing up in that world, but as that world; indeed, we are, as humans, not superior to nature, or separate from it, or above it; we are, all of us, nature. This book effectively dismantles the conservative human/nature binary.

At the same time, Blom argues that nature itself, on a global scale, is at a tipping point, yet the urgency of that is not recognised in educational systems, either by the curriculum or by those who teach them. Teachers are caught up in delivering the rigidly prescribed curricula, which iteratively reconstitutes the normativities of the status quo, such as thinking in terms of binaries and of hierarchies, and in terms of the competitive individualism of today's dominant, neoliberal mindset, and in terms of the taken-for granted, capitalist mindset that cannot, or will not, see the danger of greed and exploitation of the earth, and the exploitation of less powerful individuals. Blom's position is that while teachers are obliged to "tick the boxes" of the mandated curriculum, it is also their response-ability to forge a new awareness of human as nature along with new possibilities for creative action. Those creative, response-able movements in thought, open up, for the children, ways of thinking-in-being, where nature and humanity are not distinct entities, and where children as nature will flourish. And it opens for both teachers and children the possibility of, and commitment to, ethical practices – to what Barad has called ethico-onto-epistemologies.

This book is also a gift to those researchers who are on a quest to do research that incorporates the concepts of post-humanism and post-qualitative inquiry, yet find themselves bound to, or dependent on, qualitative methodologies before "the post" … Emergent methodologies should not be thought of in terms of linear progression, Blom argues, where the new supplants the old, drawing on Barad, who argued that we are always already haunted by the past and the future, so that neither the past nor the future can be closed. Blom offers, in place of post-qualitative methodology, the concept and practice of transqualitative inquiry, which offers researchers greater flexibility; they can draw on the well-known established practices of qualitative inquiry, while experimenting with new concepts and practices, through which human as nature can be developed.

Transqualitative methodology, as Blom explores it in this book, opens possibilities of thinking about, and knowing how to be, responsive to the power of nature. Blom listens to teachers and invites them to explore their own, and the children's, capacity for response-ability. In such work, outdoor experiences are an enactment of returning the human body to a place where it is naturally from, and to which it belongs. The findings from this study suggest that teachers' perceptions of nature (generally still) come from a human-centric position, despite any interest they might have in posthuman ideas. Nature is perceived as something external, "out-there", and as a resource: not in a destructive way, but as a place to appreciate and recharge in. What Blom offers here is a careful and detailed exploration of how it could be otherwise – more response-able, and more ethical.

Bronwyn Davies,
Professorial Fellow, University of Melbourne
Emeritus Professor, Western Sydney University

ACKNOWLEDGEMENTS

A huge thank you to Prof. Karin Murris who has been a big sup-porter, encourager, and believer of my work. It is her support that has made this book possible. Moreover, her inspirational work was pivotal in enabling me to conduct my research. The depth and quality of the research presented in this book would not be what it is without her generative work coming before. Moreover, without creating this book series, which promotes creative and innovative research methodologies, I am not certain my work would have found a place to land. Thank you Karin.

I would also like to acknowledge the incredible support of Prof. Amy Cutter-Mackenzie-Knowles and Prof. Lexi Lazcsik who guided me on this research journey and assisted in shaping and refining the ideas presented here. The work presented in this book would not have been possible without their ongoing feedback, encouragement and knowing that I could do it.

I also give enormous thank you to the participant, Caitlyn, pre-sented in this research. Without her it would not have been pos-sible. Caitlyn was so generous, enthusiastic, and insightful in all that she shared. I could not have done what I did without your thoughtful responses and amazing classroom practices. Thank you a million times over.

To my mum, Jill Lewis, who was always so proud of me no matter what I did. You taught and showed me more than you ever accepted knowing. It was immense. Thank you.

Finally, a huge thank you to my family – both my extended family and friends who are family that enrich my life everyday beyond measure – and those that I am blessed to have in my life on a daily basis. My out-of-this-world gorgeous children, Xanthia and Hudson. My husband, Joost, your tenderness, and steadiness a constant gift and rudder that holds us and keeps us living with and as the stars. I love you dearly and immeasurably.

LIST OF ACRONYMS AND ABBREVIATIONS

DET   Department of Education and Training
EYLF   Early Years' Learning Framework
IUCN   International Union for the Conservation of Nature
NSW   New South Wales
OECD   Organisation for Economic Co-Operation and
           Development
UNESCO United Nations Educational, Scientific and Cultural
           Organisation
WIRES Wildlife Information Rescue and Education Service

1

THE FIRST TURN

Thinking with/as nature for educational futures

Acknowledgement of Country

Jingi walla blagganmirr Widjabul na Jogun ba la

This means welcome to Country in the Bundjalung nation where this book was written. *I respectfully acknowledge and pay respect to the Widjabul Wia-bal Peoples of the Bundjalung Nation, the traditional owners of the beauty-full lands which this book was generated. I also respectfully acknowledge and pay respect to the Gumbaynggirr Peoples where part of this book took place.*

Knowing that this book will travel beyond the bounds of this Country, across Australia and the world, I would like to further acknowledge the Traditional Custodians of Country throughout Australia and the world and their connections to land, sea, and community. I pay my respect to Elders past, present, and emerging and extend that respect to all Aboriginal Peoples who are reading this book. May we come together, and work together, and return to our future, and our past, together.

DOI: 10.4324/9781032703473-1

Prelude

When I reflect on my childhood,
it seemed to be quite a normal experience
given the spacetime[1]:
suburban Melbourne,
Australia in the 1980s.
My memories are of family time in the local park,
school recesses and lunchtimes
in the pre-legislative adventurous natural spaces
of the school playground,
and of playing hide-n-seek
on my grandparents' boutique vineyard,
on the outskirts of the city boundary.

There were lots of opportunities for me to be in nature (see
Figure 1.1).

to "quieten the noise of city life" –
the constant barrage of distractions offered
by the electronic and other busyness
that the cityscape offered.

As teenage years encroached,
So did suburban development.
The tree-scape outside my bedroom window
was replaced
with a brick wall.
Single-house blocks into
two,
three,
and four
townhouses (see Figure 1.2).

FIGURE 1.1   Childhood with/in nature. *Photograph by Patricia Lewis. Used with permission.*

FIGURE 1.2   The "new" early 1990s view from my bedroom window. *Photograph by Simone Blom.*

Devastation hit as
my reprieve was revoked
and the escape from a somewhat
tumultuous "inside" home life
was taken away.

Global environmental catastrophes abound. These distressing calls from nature are an urgent and painful sign of human impact on the natural world.[2] Recent world happenings have demonstrated the unrelenting power of nature through record-breaking flooding, bushfires, droughts, severe storm events and viral pandemics, amongst many others, resulting in an unprecedented loss of human and nonhuman lives alike and leaving very few, if any, untouched. While the vast majority are waking up to the lack of leadership in instrumenting the urgent change that is required, the feelings of helplessness often end in quick-fix solutions to return to "business-as-usual". However, there is no going back to a normal that once was, as the Earth and times have changed. As a global collective, it is time for humans to respond to the current situation, the "new normal". With cries that the environment has been stretched beyond tipping point and the latest Organisation for Economic Co-operation and Development (OECD) and Australian Department of Education and Training (DET) reports signalling a desperate and urgent need for monumental changes to education systems.

Education, and more specifically, environmental education offers an apt avenue for learning with these unrelenting planetary warnings. Emergent environmental education philosophies and theories provide an opportune and exciting way forward for pedagogies that respond to the current planetary crisis such as research that explores posthuman concepts in environmental education is significant and timely. Moreover, research that explores emergent approaches to undertaking educational research such as transqualitative inquiry using diffractive ethnography propels environmental education into radical options for future possibilities.

This prelude explains and forms the context for this publication, both personally and planetary. A global ecological system in crisis. An education system that may offer a means to enact change but that also needs urgent disruption and transformation. Yet, a frontline of teachers who hold humongous responsibility and the weight of opportunity and possibility, are bound by the rigidity of curriculum frameworks that censor what is known and not known and an impossible volume of administration and accreditation requirements. What may responsibility mean for teachers in this current ecological scape that is inextricably socio-ecological? How might posthuman theories attend to these complexities with radical new insights to propose fresh and radical educational futures?

Introduction

My suburban upbringing
was disrupted by an 8-week camp
in a secluded National Park (see Figure 1.3).
where I reconnected with nonhuman nature –
large open spaces of green,
walks through the dense Australian bush,
accepting the responsibilities for taking care of me
and human/nonhuman others
that come with living self-sufficiently
and basking in the wonder and awe
of an endless night sky of stars.

FIGURE 1.3   Privileged opportunity in wild nature. *Photograph by Jo Wilson. Reprinted with permission.*

This pivotal experience ignited a passion
that was directed towards the environment,
then education,
then research,
to explore how environmental education
exists in early school years teachers'
perceptions and pedagogies.

In this book, I explore early school years teachers' perceptions of nature and how it informs their pedagogy. The purpose of doing so is to position educational possibilities and futures within the current practice. If transformation is needed, and I argue that it is, it requires footing in a deep understanding of what teachers currently perceive about nature and how it influences their pedagogical practice. Shifting current educational paradigms and practices requires a rigorous and robust rationale. In particular, a consideration, realisation, and recognition of the importance of educational theories for enacting future change in education. There is a need to explore and adopt theories and approaches to research and practice that challenge and change conventional ways of thinking about and enacting educational practices. These claims are not new; educational reforms have been around as long as education itself. However, these times are unlike anything that has come before and they require a different approach to investigate and instigate change. This book provides one such approach that is theoretically and methodically creative, innovative, and robust.

My motivation for writing this book is the culmination of childhood, adolescence, and adult events including the acute awareness of the preciousness of my opportunity on the eight-week "school camp", my admiration of teachers and their work and a passionate interest in the emergence and possibility of posthumanism for environmental education and its research. More specifically, I was curious about how teachers were practising environmental education in their classrooms – to see where nature "grows", how their perceptions of nature inform this practice and how posthuman theories and philosophies frame teachers' pedagogies to understand and transform current conceptualisations of human/nonhuman[3] nature relations in the classroom.

Systems in turmoil

It is no secret that the Earth as a system is facing a planetary crisis. This crisis has been labelled by some as the Anthropocene (Crutzen & Stoermer, 2000) to describe a new geographical epoch that the OECD stated signifies, "unprecedented challenges – social, economic and environmental – driven by accelerating globalisation and a faster rate of technological developments" (Costa et al., 2018, n.p.). However, at the same time I use it, I trouble the term Anthropocene for the way it is drenched in colonised thinking and as such, silences the voice of First Nations Peoples (Whyte, 2017). The planetary shifts that are now occurring are referred to as the current situation or allude more directly to the global urgencies in a move towards including all peoples in the future narrative and in an attempt to discontinue practices that marginalise and perpetuate binary and discriminatory thinking. These realities do not take away that the assertions of planetary catastrophes are real and justified, creating a culture of fear that can lead to individuals feeling disempowered and unable to make a difference (Payne, 1997; Shapiro, 2018; Sobel, 1996). Students and teachers alike are at the mercy of an educational system that does not prioritise learning with/as nature.

So far, this Turn has emphasised that the environment – nonhuman nature – is in trouble, making it known through confronting, often heart-wrenching, and undeniable happenings. These so-called extreme weather events are a devastating call out from a voice that has too long been silenced and is screaming for abnormal, normalised approaches to halt. There is no longer a normal way of life, and as a collective of human and nonhuman, the planetary cries signify the need for urgent, monumental, and radical change.

Environmental education and its research offer a suitable space to address the dire ecological predicament of the current planetary crises through encouraging and promoting practices and pedagogies that facilitate environmental awareness, knowledge, understanding, and care. Taylor (2017) argues that traditional approaches to environmental education and its research have primarily focused on the nature "out-there" and as such, have not reconciled the planetary environmental issues, despite the wealth and variety of practices these approaches have afforded and the changes they have actioned.

Moreover, social fears about uncertain futures are growing in Australia by the latest report produced by the DET which announced that "academic performance has declined when compared to other OECD countries" (Department of Education and Training [DET], 2018, p. viii). Educational reforms seeking to address this declining student performance over the last two decades (Department of Education and Training [DET], 2018, p. 2) have been largely problematic given the neoliberalist paradigm that the education system operates within (Klees, 2019). Moreover, education systems are stable and robust and therefore, by way of their very structure, resistant to change (Eisner, 1992). Eisner (1992) troubles this resistance by proposing that educational reforms consider the intentions of education, highlighting that it is not just students' test results that are important. This notion has been echoed by Weston (1996, 2004) and more recently, Blenkinsop et al. (2020), who have both challenged the foundations of education as purported through traditional school-based approaches to consider future potentialities and possibilities for education, and more specifically, environmental education.

Teachers' perceptions and pedagogies

At the coalface, teachers hold a pivotal role in inspiring children and young people to follow life paths in environmental fields (Chawla, 1998; Quinn et al., 2016; Wells & Lekies, 2006), and therefore, also enact responsibility in the current situation. Listening to the voices of teachers is imperative in response to research which has identified that the work of teachers is largely undervalued as "educational policy and restructuring have been done *to* teachers rather than *with* them. In the process, their professionalism has been denied" (Lingard et al., 2003, p. 403; italics in original). In short, the voices of teachers have traditionally been marginalised in educational research (Cutter-Mackenzie et al., 2014; Lingard & Renshaw, 2013; Lingard et al., 2003) and absent in many of the conversations that should involve and embrace their views (Bahr & Mellor, 2016). Through teachers' authentic participation in research, the significance of their contribution to the validity and integrity of the research is acknowledged. The voices of teachers are integral in providing authenticity to research that involves them and gives privilege to their thoughts, feelings, concerns, and ideas that have for too long been silenced.

The specific focus on early school years teachers draws on findings from the research literature where it has been identified that in environmental education research, "early years education and research has a vital role to play" in grappling with "an escalating intergenerational problem about how the relations of human and non-human natures are (re)configured" (Payne, 2018, p. 121). This pennant signals a need to conduct research in the early years of schooling as it provides a foundational platform from which to explore global environmental problems that are of concern to all beings (human and nonhuman alike), regardless of age. There has been a plethora of intentional teaching and many pedagogical practices are engaged in by educators in the preschool educational setting (Cutter-Mackenzie & Edwards, 2013a, 2013b; Cutter-Mackenzie & Edwards, 2014; Cutter-Mackenzie et al., 2014; Cutter-Mackenzie-Knowles et al., 2021;

Davis, 2010; Elliot & Davis, 2020). However, there is a dearth of research exploring how early school years teachers' perceptions of nature inform pedagogy in the early years of schooling when teachers are required to address sustainability curriculum priorities and school-based pedagogies are initially introduced.

In this book, the concept of "perception" casts a wide net by including previous studies that have interchangeably described teachers' perceptions in the context of conceptualisations, perspectives, attitudes, beliefs, difficulties, and self-efficacy (Cutter-Mackenzie & Smith, 2003; Fraser et al., 2015; Thi To Khuyen et al., 2020). Perception is also informed by cultural aspects including values, worldviews and social organisational structures (Qiong, 2017).

Understanding teachers' perceptions of nature is significant given existing repeated anecdotal claims about children's declining nature experiences (for example, see Louv, 2006). There has been very limited concentrated research that has turned to early school years teachers' perceptions of nature and how this informs their pedagogy through a posthuman framing.

What, why, how ... posthuman!

The burgeoning of posthuman theories in the environmental education literature (Änggård, 2016; Braidotti & Dolphijn, 2017; Crinall, 2017; Cutter-Mackenzie-Knowles et al., 2020; Malone, 2018; Murris, 2020; Rautio, 2013; Verlie, 2018) offers great hope and potential in addressing what is often referred to as a "doomsday future" (Strife, 2010). Posthumanism offers tangible and practical possibilities to combat the urgency of the present turbulent times and enables tuning in to the voices of teachers, and nature (Murris & Haynes, 2018).

Posthumanism provides a fitting platform to explore early school years teachers' perceptions of nature and how these inform their pedagogy.

Posthuman traditions trouble normalised approaches to environmental education and are essential in "acknowledging the interdependence of social and environmental issues and the importance of avoiding anthropocentric attitudes towards the environment" (Taylor, 2017, p. 1449). Further to this, posthumanism disrupts the championing of humans' incessant need to look after the nature "out-there" by identifying that "we are nature already" (Rautio, 2013, p. 394), our human body-*is*-nature (Blom, 2020; Dickinson, 2013) such that "we understand ourselves as part of nature" (Barad, 2007, p. 341).

Posthumanism provides a fitting platform for environmental education research by offering novel perspectives in addressing global environmental issues.[4] As such, posthuman theories have been gaining momentum in environmental education over the last decade (Änggård, 2016; Braidotti, 2013; Cutter-Mackenzie-Knowles et al., 2020; Malone, 2016; Murris, 2020) and their emergence is calling for new ways of thinking to be adopted in the field (Cutter-Mackenzie & Edwards, 2014; Malone, 2018, 2020; Stevenson et al., 2016). To date, empirical posthuman research in education has focused on preschool settings (Elfström Pettersson, 2017; Hackett & Somerville, 2017; Lindgren, 2020; Myers, 2020), higher education settings (Taylor, 2013), and specific educational practices such as play-based activities (Änggård, 2016; Cutter-Mackenzie-Knowles et al., 2021), literacy learning (Murris & Haynes, 2018) or specific fields such as climate change research (Cutter-Mackenzie-Knowles & Rousell, 2020; Verlie, 2018) and human-animal relations (Gannon,

2017; Malone, 2016; Myers, 2020; Taylor & Giugni, 2012; Young & Bone, 2020).

As the past research signifies, there is a critical gap and thus, a need for posthuman research in the early years of schooling in environmental education. Research studies in the early years of schooling are particularly pertinent as they enable the early intervention of environmental education and its research (Payne, 2018). Moreover, Payne (2018) identifies the dire need for research that includes the everyday experiences of young school children in various contexts such as the classroom and school environments. This research sought to meet this dearth by exploring the perceptions of nature from the classroom teacher and how this informs their pedagogy.

Malone et al. (2020, p. 3), further explore posthumanism and how it can be harnessed to reconceptualise how teachers perceive their classrooms by,

> Shifting away from the child in nature as the only agential body and focusing on the materiality of child bodies and the bodies of other nonhuman entities as relational assemblages to allow a new ethical and theoretical imagining for children and their encounters with place and nature.

The study presented in this book adopted this posthuman positioning by exploring the relationality of matter rather than human exceptionalism (for a more complete exploration of posthuman thinking, see the Third Turn). This allows for "new ethical and theoretical imagining[s]" (Malone et al., 2020, p. 3) to be explored in understanding what early school years teachers' perceptions of nature are and how they inform their pedagogy.

Transqualitative inquiry and diffractive ethnography

Leading environmental educator and researcher, Hart (2018) alerts environmental educators and researchers to the importance of reading widely outside the field to consider new ways of doing research. This book sought to respond to this call by introducing transqualitative inquiry as a methodology that enables creative research to occur within more traditional frameworks, such as PhD research, that embraces the qualitative while exploring "new" ways of working with methods (see the Fourth Turn).

Transqualitative research offers a different direction for research that challenges qualitative research frameworks and problematises post-qualitative inquiry. This proposed new offering is designed to be responsive to the current dynamic research space; where research traditions found practice but are also reconceptualised in this spacetime. This aligns research as responsive to the changes being experienced globally across social, economic, political, and environmental domains that are indeed unprecedented and call for approaches to research that challenge and re-challenge what is considered valid research. The research presented in this book acknowledges alternative research methodologies were required in response to our current spacetime.

Diffractive ethnography as an approach to transqualitative inquiry shifts thinking away from traditional ethnographic inquiry "describing people and the things they do" to asking questions about the nature of entanglements and "how boundaries are configured and reconfigured" (Gullion, 2018, p. 121). What sets diffractive ethnography apart from its foundations in conventional ethnography is that it does not focus on studying and analysing people, cultures, or groups, rather it seeks to explore phenomena (Gullion, 2018) that include relationality with human and nonhuman alike. Thinking through diffractive ethnography follows posthuman research traditions where phenomena are studied in their multitude of complex relations (Ulmer, 2017).

Moreover, the transqualitative methodology and diffractive ethnography methodological approach aligns with the concepts in this book as described through the posthuman theoretical framework (see the Third Turn) more fully than a qualitative or post-qualitative inquiry. Engaging emergent theories and methodologies in the early school years milieu provides an alternative way of viewing and understanding teachers' perceptions of nature and how these inform their pedagogy.

Outline

This book is presented in ten chapters described as a series of Turns. The word Turn is used commonly to explain a movement toward a different and "new" way of thinking. In particular, the posthuman turn, the affective turn, and the material turn are frequently described in educational and other social science research. Barad (2007) uses the concept of *returning* to describe "iteratively intra-acting, re-diffracting, diffracting anew, in the making of new temporalities (spacetimematterings), new diffraction patterns" (p. 168). In addition, this book was conceptualised through this diffractive thinking and practice, and the conceptual work is presented through the entanglement of diffractive patterns (see Figure 1.4).

The First Turn has outlined the background to the research in this book, introduced the posthuman theoretical framework, the methodological approach through diffractive ethnography and transqualitative inquiry and discussed the significance of this research in contributing knowledge to the field of environmental education.

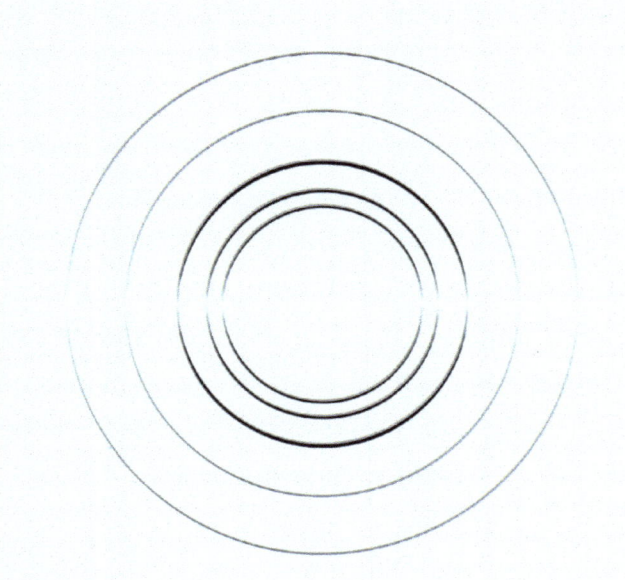

FIGURE 1.4 Diffractive patterns as portrayed in this book. *Illustration by Kelly @ Kelly Designs for the author.*

The Second Turn is the review and critical analysis of the literature that this book is founded upon. It is a returning to the past – as Barad (2007) describes – a turning over and over again like an earthworm to explain, understand, and situate this book in the present; present/ing the past for the present and future. The Third Turn presents the posthuman theoretical framework that is grounded in posthuman theory and works with the concepts of childhoodnature, affective atmospheres and Barad's (2007) concept of material-discursive practices drawn from their theory of agential realism. The Fourth Turn positions the research in this book through transqualitative methodology. The Fifth Turn describes the diffractive ethnographic approach that I adopted to undertake this research. The Sixth Turn outlines the research design and protocol for undertaking a diffractive ethnography. The Seventh Turn presents the context of spacetime underpinning the research in this book. The Eighth and Ninth Turns present the data in two parts where I have co-made diffractive data entanglements with the teacher participant to explore their perception of nature and how this informs their pedagogy. The Tenth Turn is the final Turn and presents a synthesis of the data through the literature and the theory. The book completes with an explanation of the future implications of this research.

Conclusion

Environmental education and its research are at the forefront of educational research in engaging innovative posthuman educational theories and practices. This Turn has highlighted the need for environmental education research specifically in the early school years, and also research that adopts creative and innovative theoretical and methodological approaches. Moreover, it has been identified that studies that authentically include the voices of teachers as experts in the research that concerns them are also required in the field. The dearth of environmental education research in the critical early school years requires further work to be undertaken with posthuman theorising and in the everyday contexts of school lives.

Environmental education and its research are well placed to provoke socio-ecological[5] action and change in response to the global (em)urgencies. Posthuman traditions provide a place to consider contemporary early school years teachers' perceptions of nature and how these inform their pedagogy in the current planetary crisis. This research sought to contribute to knowledge that is significant, robust, and needed in the field and more widely, to understand the potential of posthuman theories in environmental education in greater depth.

This book proposes a turn, or more accurately, a return. Not a return to the past, nor a turn to the future. This return is engaging the present through posthuman theories to understand different ways that environmental education is taking place in classrooms. In doing so, this research was designed to inform:

a. future pedagogical approaches and directions for early school years' teachers in environmental education;
b. policy and curriculum documents pertaining to sustainability and environmental education at a state and national level;
c. creative and innovative methodological practices in environmental education research, and educational research more broadly; and,
d. the application of emergent theoretical perspectives in practice.

Furthermore, this book explores the reconceptualisation of nature experiences in the contemporary school learning landscape using posthuman concepts, particularly given: (a) the time pressures of busy home lives and schedules (Malone, 2007; Malone & Tranter, 2003; Shaw et al., 2015; Sobel, 2008); (b) rigid curriculum demands (Dyment et al., 2015); and, (c) restricted access to wild nature spaces (Kahn & Hasbach, 2013; Wells & Lekies, 2006).

The purpose, and thus significance, of this book is to bring awareness to this need in a humble yet valuable contribution across four domains where there are critical gaps in the existing research. Firstly, this book makes a unique contribution to knowledge in environmental education by investigating early school years teachers' perceptions of nature and how this informs their pedagogy. Secondly, this research makes meaningful considerations to the inclusion of environmental education through Sustainability in the Australian Curriculum. Thirdly, the book introduced transqualitative inquiry as methodology using a diffractive ethnographic approach, that finally, aligned with a conceptually informed, posthuman framework. This research is significant because there is currently limited research that explores early school years teachers' perceptions of nature from this theoretical and methodological approach. There is a dire need for research that addresses these current shortfalls; both the field of education and the planet are dependent on it.

Notes

1 Spacetime is a concept first proposed by Albert Einstein, and later proven by Quantum mechanics that is used to describe mark-making practices (Barad, 2007) and the way space, and time, cannot be considered as separate entities as they are "mutually constituted through the dynamics of iterative intra-activity" (Barad, 2007, p. 181). Moreover, Barad (2007) describes how the "spacetime manifold is iteratively (re)configured in terms of how material-discursive practices come to matter. The dynamics of enfolding involve the reconfiguring of the connectivity of the spacetime matter manifold itself (a changing topology), rather than mere changes in the shape or the size of a bounded domain (geometrical shifts)" (p. 181). The dynamic way that the world makes itself known does not happen in time and space but "*in the making* of spacetime itself" (Barad, 2007, p. 140; emphasis added). That is, spacetime is the result of quantum entanglements, not an inert background where and when something takes place.
2 I acknowledge that human impact here is defined as post-industrial revolution impact by the minority world, which did not occur through the actions of First Nations Peoples.
3 I acknowledge the problematic nature of, and limitations of, language here as human/nonhuman does depict a binary yet the post-human positioning of this work contends any separatism. Murris (2016) theorises through Barad's (2007) quantum physics, feminist

positioning to trouble words as *representations* and thus, *interpretations* where the word is just a representation of something, not the actual thing. Murris (2016) argues that these representations are problematic as they are "how we think about knowledge and the relationships we build with ourselves and nonhuman others. They inform our pedagogical theories and practices and our claims of knowledge about child and childhood" (p. 45). As such, I accept these limitations and impositions of language as representations and move forward using them with a deep knowing of the problems that they unintentionally create and perpetuate.

4 Posthuman theory is explored in greater depth in the Third Turn. See the Second Turn for a critical review of the environmental education research.

5 Early socio-ecological models stemmed from ecological psychology (Lewin, 1936) and proposed that the natural environment could be viewed in two ways: anthropocentrically, as a resource for human use; or ecocentrically, where the natural world holds intrinsic worth regardless of the perceived importance to humans (Eckersley, 1992; O'Riordan, 1976; Sessions, 1974). More recent socio-ecological models emphasise the significance of the relationship and interactions between humans and the natural environment (Wattchow et al., 2013).

References

Änggård, E. (2016). How matter comes to matter in children's nature play: Posthumanist approaches and children's geographies. *Children's Geographies*, *14*(1), 77–90. https://doi.org/10.1080/147332 85.2015.1004523

Bahr, N., & Mellor, S. (2016). *Building quality in teaching and teacher education* (Vol. 61). Australian Council for Educational Research.

Barad, K. (2007). *Meeting the universe halfway: Quantum physics and the entanglement of matter and meaning*. Duke University Press.

Blenkinsop, S., Jickling, B., Morse, M., & Jensen, A. (2020). Wild pedagogies: Six touchstones for childhoodnature theory and practice. In A. Cutter-Mackenzie, K. Malone, & E. Barratt Hacking (Eds.),

Research handbook on childhoodnature: Assemblages of childhood and nature research (pp. 1–18). Springer International Publishing.

Blom, S. M. (2020). Conceptualizing parent(ing) childhoodnature through significant life experience. In A. Cutter-Mackenzie, K. Malone, & E. Barratt Hacking (Eds.), *Research handbook on childhoodnature: Assemblages of childhood and nature research* (pp. 1–26). Springer International Publishing. https://doi.org/10.1007/978-3-319-51949-4_127-1

Braidotti, R. (2013). *The posthuman*. Polity Press.

Braidotti, R., & Dolphijn, R. (2017). *Philosophy after nature*. Rowman & Littlefield International.

Chawla, L. (1998). Significant life experiences revisited: A review of research. *Journal of Environmental Education, 29*(3), 11. http://ezproxy.scu.edu.au/login?url=http://search.ebscohost.com/login.aspx?direct=true&db=pbh&AN=340880&site=ehost-live

Costa, J., Dillon, S., Suzuki, K. H., Kim, M., Skovsgaard, J., & Schleicher, A. (2018). *Education for a better world: An ongoing debate at a global scale*. https://www.oecd.org/education/school/Education-for-a-better-world-an-ongoing-debate-at-a-global-scale.pdf

Crinall, S. (2017). Body-place-time: Painting and blogging "dirty, messy" human-natured becomings. In K. Malone, S. Truong, & T. Gray (Eds.), *Reimagining sustainability in precarious times* (pp. 95–114). Springer Singapore. https://doi.org/10.1007/978-981-10-2550-1_7

Crutzen, P. J., & Stoermer, E. F. (2000). The "Anthropocene". *Global Change Newsletter, 41*, 17–18.

Cutter-Mackenzie, A., & Edwards, S. (2013a). The next 20 years: Imagining and re-imagining sustainability, environment and education in early childhood education. In S. Elliott, S. Edwards, J. Davis, & A. Cutter-Mackenzie (Eds.), *Early childhood Australia's best of sustainability: Research, practice and theory* (pp. 61–67). Early Childhood Australia.

Cutter-Mackenzie, A., & Edwards, S. (2013b). Toward a model for early childhood environmental education: Foregrounding, developing, and connecting knowledge through play-based learning. *The Journal of Environmental Education, 44*(3), 195–213. https://doi.org/10.1080/00958964.2012.751892

Cutter-Mackenzie, A., & Edwards, S. (2014). Everyday environmental education experiences: The role of content in early childhood education. *Australian Journal of Environmental Education, 30*(1), 127.

Cutter-Mackenzie, A., Edwards, S., Moore, D., & Boyd, W. (2014). *Young children's play and environmental education in early childhood education*. Springer. https://doi.org/10.1007/978-3-319-03740-0

Cutter-Mackenzie-Knowles, A., Malone, K., & Barratt Hacking, E. (2020). *Research handbook on childhoodnature: Assemblages of childhood and nature research* (A. Cutter-Mackenzie, K. Malone, & E. Barratt Hacking, Eds.). Springer International Publishing.

Cutter-Mackenzie-Knowles, A., Osborn, M., Lasczik, A., Malone, K., & Knight, L. (2021). *The Mudbook: Nature play framework*. Queensland Government Department of Education.

Cutter-Mackenzie-Knowles, A., & Rousell, D. (2020). The mesh of playing, theorizing, and researching in the reality of climate change: Creating the co-research playspace. In A. Cutter-Mackenzie, K. Malone, & E. Barratt Hacking (Eds.), *Research handbook on childhoodnature*. Springer International Handbooks of Education.

Cutter-Mackenzie, A., & Smith, R. (2003). Ecological literacy: The "missing paradigm" in environmental education (part one). *Environmental Education Research, 9*(4), 497–524.

Davis, J. (2010). What is early childhood education for sustainability? In J. Davis (Ed.), *Young children and the environment: Early learning for sustainability* (pp. 21–42). Cambridge University Press.

Department of Education and Training [DET]. (2018). *Through growth to achievement: Report of the review to achieve educational excellence in Australian schools*. Commonwealth of Australia.

Dickinson, E. (2013). The misdiagnosis: Rethinking "nature-deficit disorder". *Environmental Communication: A Journal of Nature and Culture, 7*(3), 315–335.

Dyment, J., Hill, A., & Emery, S. (2015). Sustainability as a cross-curricular priority in the Australian curriculum: A Tasmanian investigation. *Environmental Education Research, 21*(8), 1105–1126.

Eckersley, R. (1992). *Environmentalism and political theory, towards an ecocentric approach*. State University of New York Press.

Eisner, E. W. (1992). Educational reform and the ecology of schooling. *Teachers College Record, 93*(4), 610–627.

Elfström Pettersson, K. (2017). Teachers' actions and children's interests: Quality becomings in preschool documentation. *Nordisk Barnehageforskning, 14*(2), 1–17. https://doi.org/10.7577/nbf.1756

Elliot, S., & Davis, J. (2020). Challenging taken-for-granted ideas in early childhood education: A critique of Bronfenbrenner's ecological systems theory in the age of post-humanism. In A. Cutter-Mackenzie, K. Malone, & E. Barratt-Hacking (Eds.), *Research handbook on childhoodnature: Assemblages of childhood and nature research*. Springer International Publishing.

Fraser, J., Gupta, R., & Krasny, M. E. (2015). Practitioners' perspectives on the purpose of environmental education. *Environmental Education Research, 21*(5), 777–800.

Gannon, S. (2017). Saving squawk? Animal and human entanglement at the edge of the lagoon. *Environmental Education Research, 23*(1), 91–110. https://doi.org/10.1080/13504622.2015.1101752

Gullion, J. S. (2018). *Diffractive ethnography*. Routledge. https://doi.org/https://doi.org/10.4324/9781351044998

Hackett, A., & Somerville, M. (2017). Posthuman literacies: Young children moving in time, place and more-than-human worlds. *Journal of Early Childhood Literacy, 17*(3), 374–391.

Hart, P. (2018). *A 50-year retrospective on environmental education inquiry: Recognizing the past and challenging the future*. Creating Capacity for Change.

Kahn, P. H., & Hasbach, P. H. (2013). *The rediscovery of the wild*. MIT Press.

Klees, S. J. (2019). Capitalisation and global education reform. In K. J. Saltman & A. J. Means (Eds.), *The Wiley handbook of global educational reform*. John Wiley & Sons.

Lewin, K. (1936). *Principles of topological psychology*. McGraw-Hill.

Lindgren, T. (2020). The figuration of the posthuman child. *Discourse: Studies in the Cultural Politics of Education, 41*(6), 914–925.

Lingard, B., Hayes, D., & Mills, M. (2003). Teachers and productive pedagogies: Contextualising, conceptualising, utilising. *Pedagogy,*

Culture & Society, *11*(3), 399–424. https://doi.org/10.1080/14681360300200181

Lingard, B., & Renshaw, P. (2013). Teaching as a research-informed and research-informing profession. In A. Campbell & S. Groundwater-Smith (Eds.), *Connecting inquiry and professional learning in education* (pp. 40–53). Routledge.

Louv, R. (2006). *Last child in the woods: Saving our children from nature-deficit disorder*. Atlantic Books.

Malone, K. (2007). The bubble-wrap generation: Children growing up in walled gardens. *Environmental Education Research, 13*(4), 513–527. https://doi.org/10.1080/13504620701581612

Malone, K. (2016). Reconsidering children's encounters with nature and place using posthumanism. *Australian Journal of Environmental Education, 32*(1), 1–15. https://doi.org/10.1017/aee.2015.48

Malone, K. (2018). *Children in the Anthropocene: Rethinking sustainability and child friendliness in cities*. Palgrave Macmillan, Springer Nature.

Malone, K. (2020). Re-turning childhoodnature: A diffractive account of the past tracings of childhoodnature as a series of theoretical turns. *Research handbook on childhoodnature: Assemblages of childhood and nature research* (pp. 1–31). Springer.

Malone, K., Duhn, I., & Tesar, M. (2020). Greedy bags of childhoodnature theories. In A. Cutter-Mackenzie-Knowles, K. Malone, & E. Barratt Hacking (Eds.), *Research handbook on childhoodnature*. Springer International Publishing.

Malone, K., & Tranter, P. J. (2003). School grounds as sites for learning: Making the most of environmental opportunities. *Environmental Education Research, 9*(3), 283–303. https://doi.org/10.1080/13504620303459

Murris, K. (2016). *The posthuman child: Educational transformation through philosophy with picture books*. Routledge. https://doi.org/10.4324/9781315718002

Murris, K. (2020). Posthuman child and the diffractive teacher: Decolonizing the nature/culture binary. In A. Cutter-Mackenzie, K. Malone, & E. Barratt Hacking (Eds.), *Research handbook on childhoodnature: Assemblages of childhood and nature research* (pp. 1–25). Springer International Publishing. https://doi.org/10.1007/978-3-319-51949-4_7-2

Murris, K., & Haynes, J. (2018). *Literacies, literature and learning: Reading Classrooms differently*. Routledge.

Myers, C. Y. (2020). "I don't know what's gotten in to me, but I'm guessing It's snake germs": Becoming beasts in the early years classroom. In A. Cutter-Mackenzie, K. Malone, & E. Barratt Hacking (Eds.), *research handbook on childhoodnature: Assemblages of childhood and nature research* (pp. 1–25). Springer International Publishing. https://doi.org/10.1007/978-3-319-51949-4_71-1

O'Riordan, T. (1977). Environmental ideologies. *Environment and Planning A, 9*(1), 3–14.

Payne, P. (1997). Embodiment and environmental education. *Environmental Education Research, 3*(2), 133–153. https://doi.org/10.1080/1350462970030203

Payne, P. G. (2018). Early years education in the Anthropocene: An ecophenomenology of children's experience. *International handbook of early childhood education* (pp. 117–162). Springer.

Qiong, O. (2017). A brief introduction to perception. *Studies in Literature and Language, 15*(4), 18–28.

Quinn, F., Castéra, J., & Clément, P. (2016). Teachers' conceptions of the environment: Anthropocentrism, non-Anthropocentrism, anthropomorphism and the place of nature. *Environmental Education Research, 22*(6), 893–917. https://doi.org/10.1080/13504622.2015.1076767

Rautio, P. (2013). Children who carry stones in their pockets: On autotelic material practices in everyday life. *Children's Geographies, 11*(4), 394–408.

Sessions, G. (1974). Anthropocentrism and the environmental crisis. *Social Behaviour and Natural Environments, 2*(1), 71–81.

Shapiro, B. (2018). *Actions of their own to learn* (B. Shapiro, Ed.). Brill Sense.

Shaw, K., Anderson, D. M., & Barcelona, B. (2015). Parental perceptions of constraints to family participation in nature-based, outdoor experiences. *Journal of Outdoor Recreation, Education and Leadership, 7*, 3–19. https://go.galegroup.com.ezproxy.scu.edu.au/ps/i.

do?p=EAIM&sw=w&u=scu_au&v=2.1&it=r&id=GALE%7CA41413
5741&asid=f1d30e8951aed06f9d0f257498d7a05f

Sobel, D. (1996). *Beyond ecophobia. Reclaiming the heart in nature education.* Orion Society.

Sobel, D. (2008). *Childhood and nature: Design principles for educators.* Sternhouse Publishers.

Stevenson, R. B., Ferreira, J.-A., & Emery, S. (2016). Environmental and sustainability education research, past and future: Three perspectives from late, mid, and early career researchers. *Australian Journal of Environmental Education, 32*(1), 1–10.

Strife, S. (2010). Reflecting on environmental education: Where is our place in the green movement? *The Journal of Environmental Education, 41*(3), 179–191.

Taylor, A. (2017). Beyond stewardship: Common world pedagogies for the Anthropocene. *Environmental Education Research, 23*(10), 1448–1461. https://doi.org/10.1080/13504622.2017.1325452

Taylor, C. A. (2013). Objects, bodies and space: Gender and embodied practices of mattering in the classroom. *Gender and Education, 25*(6), 688–703. https://doi.org/10.1080/09540253.2013.834864

Taylor, A., & Giugni, M. (2012). Common worlds: Reconceptualising inclusion in early childhood communities. *Contemporary Issues in Early Childhood, 13*(2), 108–119. https://doi.org/10.2304/ciec.2012.13.2.108

Thi To Khuyen, N., Van Bien, N., Lin, P.-L., Lin, J., & Chang, C.-Y. (2020). Measuring teachers' perceptions to sustain STEM education development. *Sustainability, 12*(4), 1531. https://www.mdpi.com/2071-1050/12/4/1531

Ulmer, J. B. (2017). Posthumanism as research methodology: Inquiry in the Anthropocene. *International Journal of Qualitative Studies in Education, 30*(9), 832–848.

Verlie, B. (2018). Affective entanglements: Learning to live-with climate change [Ph.D. thesis]. Monash University.

Wattchow, B., Jeanes, R., Alfrey, L., Brown, T., Cutter-Mackenzie, A., & O'Connor, J. (2013). *The socioecological educator: A 21st century renewal of physical, health, environment and outdoor education [book].* Springer. http://ezproxy.scu.edu.au/login?url=http://search.

ebscohost.com/login.aspx?direct=true&db=nlebk&AN=637326&site=ehost-live

Wells, N. M., & Lekies, K. S. (2006). Nature and the life course: Pathways from childhoodnature experiences to adult environmentalism. *Children Youth and Environments*, *16*(1), 1–24.

Weston, A. (1996). Deschooling environmental education. *Canadian Journal of Environmental Education (CJEE)*, *1*(1), 35–46.

Weston, A. (2004). What if teaching went wild? *Canadian Journal of Environmental Education*, *9*(1), 31–46.

Whyte, K. (2017). Indigenous climate change studies: Indigenizing futures, decolonizing the Anthropocene. *English Language Notes*, *55*(1), 153–162.

Young, T., & Bone, J. (2020). Troubling intersections of childhood/animals/education: Narratives of love, life, and death. In A. Cutter-Mackenzie, K. Malone, & E. Barratt-Hacking (Eds.), *Research handbook on childhoodnature*. Springer International Publishing.

2

THE SECOND TURN: TEACHERS' PERCEPTIONS WITH/AS NATURE IN ENVIRONMENTAL EDUCATION, A RETURNING

A critical review of the literature in environmental education about teachers' perceptions and pedagogies as nature

Introduction

The purpose of this book is to explore teachers' perceptions of nature and how it informs pedagogy through a posthuman perspective. As such, in this Turn,[1] I return to the extant research literature in the field. Returning past research to the present that responds to questions of how teachers perceive nature and how it informs their pedagogy contextualises the research presented in this book within the field of environmental education and education more broadly. In addition, this Turn opens up the possibilities for future research. It asks the reader to read more deeply to see what is hiding, what remains to be seen and where we can move our gaze beyond what we know to explore the unknown. Your involvement in this Turn is not passive, but an intra-activity. Whatever seeds you find in your exploration, are yours to take and grow. In this (re)turn, the existing research literature is presented as two areas of scholarship: environmental education, and teachers' environmental perceptions and pedagogies. However, before these two areas are explored, this Turn begins

DOI: 10.4324/9781032703473-2

by problematising current school-based education in Western minority world culture.

What is the purpose of school-based education in Western minority world culture?

Education was a matter of leading novices out into the world rather than, as commonly understood today, of instilling knowledge into their minds.

Ingold (2014, p. 388)

Modern schooling in the minority, post-colonial world, was born from the theoretical and philosophical origins of classical Greek philosophers such as Pythagoras, Heraclitus, Plato, and Socrates (Stanford University, 2022; Stehlik, 2018). Modern or "mass" schooling was developed in response to capitalism: the onset of the Industrial Revolution where an increase in production capability resulted in an increased market for skilled workers. The result was an education system that was supposedly more accessible, efficient, and productive (Stehlik, 2018). However, the rise of this industrial model in society was at the cost of the environment, which continues to be given very little consideration.

There are many views on the purpose of education. Given the capitalist, political agenda that school-based education in minority cultures currently occupies, there is significant interest in increasing the economic growth of a country (Smith et al., 2016). This occurrence has been called "Human Capitalist Theory" to explain the "site of the individual" and how "her or his skills, knowledge, competencies and behaviours … contribute to the productivity of the nation or country" (Smith et al., 2016, p. 125). Standardised testing reports on the educational "health" of a nation for world economic rankings and ratings and to inform future reforms. According to standardised testing, educational reforms in Australia have not bettered academic performance nor the health and well-being of the school student population (DET, 2018) and unarguably, the health and well-being of non-human nature. Such educational reforms operate within neoliberalist economic structures; however, they are only doing so for the privileged few at the expense of the majority (Klees, 2019).

Smith et al. (2016) provide a similar assessment of the system in stating that "neoliberalism pretends to provide 'equal' and 'fair' opportunities for 'all' to achieve 'success' while it has re-inscribed, intensified and continued to create injustices and inequity" (p. 130). This occurs as capitalist neoliberal structures resist accountability by diverting attention away from the system and "casting the blame for education and development problems elsewhere … [on] individuals for their lack of 'investment' in human capital, for their not attending school, for their dropping out of school, for their not studying the 'right' fields" (Klees, 2019, p. 19). In short, the system pushes against and, in effect, blames those individuals who do not fit the model of human capital it espouses (Klees, 2019; Moore, 2016; Smith et al., 2016). However, it is no fault of teachers or other individuals as the system continues to operate through its own momentum. Despite the vital role that education holds in inciting and enacting responses to the global environmental crisis, under the neoliberal, capitalist structure described here, the reasons nature continues to suffer under the duress of human actions become clearer.

Braidotti and Dolphijn (2017, p. 6) explain:

Being so caught up in the systems of consumption, and thus unable to see the vast desertification induced by the capitalist machineries for the last two centuries, one wonders whether this clever animal has not vacated the stage for some time. Maybe the exacerbated sense of subjectivity and worldliness that capitalism continues to sell is just like one of those bright stars that we see glittering in the sky but that in reality had ceased to exist long ago.

While "seeing" the "vast desertification" may be the privilege of but a few, Braidotti and Dolphijn (2017) affirm that as a species, little has changed over the last 200 years as the cogs of capitalism continue to turn. The forces that come from this dominating world system of capitalism (Wallerstein, 2004) promote "greed, inequality, and environmental destruction" (Klees, 2019, p. 20). Moreover, capitalism utilises racism and sexism to fuel this force that is "extraordinarily resistant to change" (Klees, 2019, p. 20). Klees (2019) warns that "we need to be very cognizant of the forces arrayed against progressive change" (p. 20). This call is for greater awareness and acknowledgement of these forces. Furthermore, finding ways to operate within the education system without being a victim to these forces would provide a timely forward step for future educational reforms.

Educational reforms that operate under neoliberal economics (endorsing the wealth of some and the belittling of others) in the belief that they are supportive of effective solution generation are problematic (Klees, 2019). The challenge then becomes, what does education "look like" in consideration of nonhuman other within the constructs of this deeply ingrained social and economic fabric? Or as questioned by Braidotti and Dolphijn (2017), in the wider context, "how to open our eyes to the relations that incorporate us in the world, that connect everything but were simply left out of the equation by capitalism?" (p. 7).

It is evident that educational practices that bring awareness to the issues aroused by capitalism, particularly those that continue to neglect the environment, are desperately needed. Bringing awareness to capitalism provides a mechanism for opening our eyes, removing the sheath of darkness and thus exposing a glimpse of what relationality with/in the world actually is (Braidotti & Dolphijn, 2017).

Capitalism filters through the education system in many everyday school-based practices enforced by separatist, competitive strategies. This is the context in which environmental education is situated within. A schooling system where competition and individualism are encouraged and championed over collaboration and togetherness.

Environmental education: backgrounds, challenges, and opportunities

Environmental education as a field of study is deeply rooted in a minority-world, colonialised history. However, I acknowledge that the practices and principles that underpin environmental education are much older. First Nations Peoples and Cultures naturally tended to nature, to Country – from what many white people now consider a posthuman viewpoint – that the human is Country, they are one (Cutter-Mackenzie-Knowles et al., 2020). As such, Country was honoured and cared for by First Nations Peoples and Cultures, such that they understood the importance of sustainability long before it was conceptualised in Westernised minority world thinking (Cutter-Mackenzie-Knowles et al., 2020; Hatcher, 2012; Whitehouse et al., 2014). While much of this ancient knowledge has been lost and was not shared as it could have been into postcolonial practices, it is recognised that environmental education is not "new" when considered through the lens of the 75,000-year history that First Nations peoples have had with/as Country (Cutter-Mackenzie-Knowles et al., 2020).

From a minority-world view, early philosophers, such as Rousseau, advocated for education with and in nature (Rousseau, 1976). However, Rousseau's ideas perpetuated the dualism of nature/culture or nature/society through his thinking that nature was the "antithesis of society's shortcomings" and he also proposed a "conflation of childhood with Nature" (Taylor, 2013, pp. 6–7). Rousseau was ostracised for his ideas about children having agency, being treated equally and learning from nature, which were considered radical at the time, and he was forced to flee his home in fear of arrest and persecution (Rousseau, 1976). Rousseau's exile marks but one event in a sustained tradition of marginalisation of environmental education (Cutter-Mackenzie & Edwards, 2014; Davis & Elliott, 2003).

Education with, in, for and about nature continued as a field of practice into the 19th and early 20th centuries through the works of philosophers and pedagogues (Palmer, 2002; Stehlik, 2018) such as Thoreau (Thoreau, 2004), Froebel (Froebel, 2005), Montessori (Montessori, 1917) and Dewey (Dewey, 1916). Nature, just like childhood and science, is a human construct, and historically, humans have conceptualised nature in many ways (Cronon, 1996; Devall & Sessions, 1985; Evernden, 1992; Latour, 2004; Orr, 1992; Sheldrake, 1990; Soper, 1995). In early Greek philosophy, education with, in, for and about nature, was not separate from everyday life (Stehlik, 2018). Greek philosophers viewed nature as alive and intelligent; they proposed that the matter of nature constituted the materiality of the world, the psychical component of nature contributed to the world's life processes and the intelligibility of nature was part of the world's mind (Colling-wood, 1960). The Renaissance period positioned nature similarly in acknowledging the intelligence of nature through the order of the natural world; however, they saw the intelligence belonging to a "divine creator and ruler of nature" (Collingwood, 1960, p. 5) rather than nature itself. Modern postulations of nature accept a realist perspective of the natural world infinitely influenced by the sociocultural ideas and beliefs that humans hold and project onto their conceptions (Cronon, 1996).

Into the mid-20th century, environmental education saw an upwelling of care and concern for the natural environment, which aligned with the growing awareness of the increasing environmental impacts of industrial pollution. The works of Leopold (1949), Carson (1965), and Bookchin (1987) alerted the world to the consequences of these impacts, and their seminal works sought to increase greater awareness, agency, action, and change.

In 1970, in response to a fast-growing movement and calls from various groups and individuals, the World Conservation Union (also known as the International Union for the Conversation of Nature and Natural Resources or the IUCN) and the United Nations Educational, Scientific and Cultural Organization (UNESCO) set a historical marker. They hosted the International Working Meeting on Environmental Education in the school curriculum, where they formally defined environmental education (Dyment et al., 2015; Edwards, 2015; Gough & Gough, 2010; Palmer, 2002; Stevenson et al., 2016). The definition focussed attention towards the values, skills and attitudes needed to "understand and appreciate the inter-relatedness among man [sic], his culture, and his biophysical surroundings" (IUCN, 1970, p. 7). This act also declared the importance of critical decision-making and a "code of behaviour" in environmental issues (IUCN, 1970, p. 7). Asserting the importance and significance of environmental education through this deed gave environmental education weight, validation and currency in the international political, social and education arena. Despite identity confusion through the attempted rebranding of environmental education as sustainability of/as/for the environment, it is noted that environmental education has been acknowledged as "one of the most critical elements of an all-out attack on the world's environmental crisis" (UNESCO, 1976, p. 2). Unarguably, over fifty years on, this is still the case.

More recently, in troubling the human/nonhuman[2] nature relationship, which is an integral consideration for environmental education, Haraway (2012) heralds that "we must find another relationship to nature" (p. 296). Haraway objects to the (mis/ab) use of nature through its possession, self-centred attempts at conservation, preservation and stabilisation, especially through tourist excursions and marking off nature parks. Cronon (1996) further asserts that although there is an obvious impact of humanity on nonhuman nature, that to forget nature's dynamism, flux and entanglement with human history is a "deeply problematic assumption" (p. 24). This is not to negate the obvious impacts that human behaviours, practices and actions are placing on nonhuman nature but to acknowledge the greater fact that human bodies *are* nature. Jickling et al. (2018) argues this necessary transition in the human relationship with the natural world "from a dominantly human-centred orientation into one that is much more equitable and interactive" (p. 168). Nature is not something outside of the human; the human body in its discrete physical form *is* nature. This perspective is offered by Barad (2007) who states, "we are a part of that nature that we seek to understand" (p. 26).

Adopting this view, which deconstructs dualistic tendencies of nature/culture or human/nature, enacts the inevitable entanglement of human bodies *as* nature. This posthuman perspective is evident through such research described by Blenkinsop (2018) who outlines the importance of children experiencing "themselves as nature" (p. 8) and Blom (2020) who refers to this as the human "body-as-nature". Rautio (2013a) extends this contention by concluding that "being nature is to construe yourself as belonging to a universe that articulates through you and extends beyond you" (p. 455). Nature is not something "out-there" (Payne, 1998) but a responsibility to understand the intra-connectedness of the body of nature, to which each human body-as-nature belongs.

Since its formal conception, environmental education has progressed and transformed along with global needs and trends and continues to do so. Environmental education remains a significant contributor to developing increased awareness, agency and action towards tackling global environmental issues. Environmental education holds a noteworthy place in the global education space of, what some are calling, the "Capitalocene" (Haraway, 2015; Malm & Hornborg, 2014; Moore, 2017). This juxtaposition, of operating with/in the economic and political forces driving the education system are the same forces that create environmental catastrophes, creates a tension in the field known as a nature/society dualism (Moore, 2016) or, as described above, a nature/culture divide.

Some of the recent, notable creative and divergent thinking in environmental education is exemplified by Taylor (2017) who borrows Latour's (2004) concept of "common worlds" to propose a theoretical framework for early childhood education in response to the Anthropocene. "Common worlds" promotes the entangled human/nonhuman world and espouses a "low-key, ordinary, everyday kind of response that values and trusts the generative and recuperative powers of small and seemingly insignificant worldly relations" (p. 1459). However, Rousell and Cutter-Mackenzie-Knowles (2020) challenge this view and argue that an "uncommon worlds" aesthetic of childhood be considered. This is one that enacts a "new" politics and ethics of relationality which accepts "both the 'common world' of nature as an extensive continuum and the 'uncommon worlds' through which children make aesthetic contact with other creatures, environments, and modes of existence" (Rousell & Cutter-Mackenzie-Knowles, 2020, p. 1659). These rich theoretical and philosophical insights into reconceptualising environmental education demonstrate the creative, speculative and affective turn that researchers are exploring.

Environmental education research is currently calling for a meaningful understanding of these human/nonhuman relations with/in the context of the schooling milieu (Cutter-Mackenzie & Edwards, 2014; Stevenson et al., 2016), particularly where there is a focus on learning and learning experiences (Edwards et al., 2015; Stevenson et al., 2016). Further to this, there is the ever-present need for greater translation of theory into practice (Hart, 2010; Palmer, 2002; Robertson & Krugly-Smolska, 1997; Stevenson, 2007; Stevenson et al., 2016).

Teachers' environmental perceptions and pedagogies

The inspiration for this research and this book has always been teachers. Teachers have constantly been a point of inspiration for my work; both as a teacher myself and as a researcher. Therefore, as we move through this book exploring teachers' perceptions and pedagogies with/as nature, I turn our focus now to the voices of teachers. In doing so, I turn the spotlight on understanding how teachers perceive nature, by examining teachers' perceptions, of, with and as nature and the environment, and how this informs their pedagogy through the existing literature.

Teachers' perceptions of nature

As discussed in the First Turn, when exploring the existing research literature, the concept of "perception" is an umbrella term used to include many others. Perception describes conceptualisations, perspectives, attitudes, beliefs, difficulties and self-efficacy (Cutter-Mackenzie & Smith, 2003; Fraser et al., 2015; Thi To Khuyen et al., 2020). Perception is informed by many social and cultural aspects such as values, worldviews and social organisational structures (Qiong, 2017).

In their work with preservice teachers, Desjean-Perrotta et al. (2008) identified that preservice teachers' perceptions of the environment and level of understanding were akin to that of the children; they were very descriptive *about* the environment – as external to the human. Desjean-Perrotta et al. (2008) further reported that environmental perceptions "correlate directly to human activities related to the environment" (p. 21). However, there is an "underlying assumption" that teachers hold a level of knowledge and understanding in environmental literacy and how this informs and affects behaviour. It was identified this is not always the case as

> …conceptual diversity about the environment can result from individual experiences and what the current political or social media may be saying about this topic at any given moment.

We, therefore, cannot assume preservice teachers have a common knowledge base about the environment or a well-developed definition. Neither can we expect them to have the basic skills and content knowledge, especially of environmental processes and systems, required of environmentally literate citizens.

Desjean-Perrotta et al. (2008, p. 22)

Through the pivotal work of Cutter-Mackenzie and Smith (2003) on environmental education and in particular eco-literacy, it was identified that teachers' environmental beliefs are placed along a continuum according to four philosophies: cornucopian, accommodation, eco-socialist and Gaia. Most participants in their study identified with the eco-socialist perspective that "the environment should be protected, even if it results in a reduction in economic growth" (p. 513). The study also revealed that, in general, teachers are lacking in environmental content knowledge with the majority of participants not considering that knowledge was of importance compared to values, attitudes and actions in environmental education.

A decade later, Torquati et al. (2013) categorised their findings into seven themes after asking 162 early childhood professionals from Nebraska and South Dakota in the USA, "What is Nature?". Of these seven themes, only one of the categories suggested an alignment with the notion that human bodies are nature: "All around us/encompassing, we are all connected", which was described as "we are part of nature; nature is all around us; nature encompasses all" (Torquati et al., 2013, p. 733). However, this description is problematic as it positions humans as both "part of nature" and not a part of nature ("nature is all around us").

So, the category of "all around us/encompassing, we are all connected" is not explicitly related to humans being included in conceptualisations of nature; however, it is one understanding of nature. Other international studies have been inconclusive in identifying whether teachers perceive themselves *as* nature (Flogaitis & Agelidou, 2003; García-González et al., 2020; Kimaryo, 2011; Maurer & Bogner, 2020; Munoz et al., 2009).

For example, Kimaryo (2011) interviewed primary school teachers in Tanzania about their perceptions of the environment in two phases: the first phase asked what the environment means to them and the second phase asked if man [*sic*] was part of the environment. However, this point was not elaborated upon, and the research findings were contextualised in the traditional education *in, for* and *about* the environment without extrapolating further on education *as* the environment.

In a European study that surveyed the environmental attitudes of 6,379 pre- and in-service teachers from 16 European countries, it was identified that teachers' attitudes towards the environment could be considered as two-dimensional: either Preservation or Utilitarian (Munoz et al., 2009), with no option for the inclusion of the teacher as part of the environment they were being surveyed about. The dimension that the teacher was identified within was correlated with their socio-economic situation, and their discipline in teaching and influenced their level of environmental involvement at the local, association and professional levels (Munoz et al., 2009).

In a study in Athens, kindergarten teachers' conceptions about the environment and nature were identified in a qualitative study to be "naturalistic, simplistic, limited and enriched with romantic elements" while the environment focused on the "biophysical dimension" (Flogaitis & Agelidou, 2003, p. 475). Similar findings came from a study by Maurer and Bogner (2020) that investigated first-year Swiss-German university students' perceptions of nature and the environment. Again, using a qualitative methodology, they found that participants identified nature with positive feelings and emotions such as calm, joy and aesthetic appreciation while considering humans to be the greatest environmental threat. This demonstrates a very human-centred view and further findings from this study, stated that "humans" did not inform participants' perception of nature (Maurer & Bogner, 2020).

In Spain, García-González et al. (2020) sought to understand pre-service teachers' perceptions and knowledge of Education for Sustainability by adopting a traditional mixed-methods approach of quantitative and qualitative research methodologies. The study involved an intervention to assess pre-service teachers' perceptions and knowledge prior to and post the intervention. Of interest were the pre-perceptions of the pre-service teachers as falling into the three categories of (i) a vague answer that does not situate sustainability within education; (ii) a perception for the conservation of the environment and natural resources (naturalist position); and (iii) a perception of the natural environment and social or economic improvement without integrating the three of them. There were no participants who were initially categorised into the two more complex levels as including all the spheres or to improve the quality of life. However, there was no opportunity for participants to demonstrate a perception *as* nature (García-González et al. (2020).

An Australian study of teachers' conceptions of the environment was analysed by Quinn et al. (2016) through the concepts of anthropocentrism and anthropomorphism. They found that positive attitudes towards nature and the environment were shown in the case of teachers with non-anthropocentric and anthropomorphic attitudes, who were also described as eco-literate educators. Moreover, the results were consistent with those from Cutter-Mackenzie and Smith (2003) such that "both point to diversity in perspectives of teachers in relation to anthropocentrism/non-anthropocentrism, with sizeable anthropocentric representation" (p. 13). In another Australian study, Evans et al. (2012) found that in a study of 30 pre-service teachers, they understood EfS according to four categories: "(1) education that is continuous (long term); (2) education about ecological systems and environmental issues; (3) education that is active, hands-on, local and relevant; and (4) education for the future" (p. 5). These categories were all situated in the traditional three-pronged approach of education for, of and about the environment and sustainability to varying extents.

Ernst (2014) highlighted that there is "mixed evidence regarding the extent of influence of teacher beliefs on their practice" (p. 739). However, despite and because of this, investigating teachers' perceptions (which includes beliefs, as per the discussion earlier in this section) is imperative to understand what informs teachers' practices and their professional learning needs. Extensive and long-standing literature exists on the relationship between teachers' beliefs and their practices generally (Aguirre & Speer, 1999; Calderhead, 1996; Clark & Peterson, 1984; Waters-Adams, 2006), including the work by Evans et al. (2012) which confirms that "teachers have views that guide and interact with their practice" (p. 2), and that "what teachers know, think and believe directly affects classroom content and pedagogy" (p. 3). However, Ernst (2014) highlights that this has not been largely applied to the environmental education context, and from my literature analysis, nor from a posthuman theoretical positioning.

Interestingly, Fägerstam (2012) conducted a study that purported to investigate teachers' perceptions of children and young people's experience of nature; however, the study focussed more on how children experienced nature than a study into the psychology of teachers' perceptions. Moreover, it did not identify the teachers' perceptions of nature and how this might have informed their perceptions of the observations of their students.

From this concise overview of the extant research literature, it is highlighted that teachers' perceptions have been sought previously; however, how they have been sought has been within a framework that has not considered or allowed perceptions outside traditional ideas of *in*, *for*, *about* and *with* the environment. Research has not provided the opportunity for teachers to share their perceptions of their bodies being *as* nature, and as such their views and perceptions have been somewhat censored.

Teachers' pedagogies in environmental education

In environmental education pedagogy, the teacher plays a significant role in a student's ability to engage with and become interested in environmental issues and care, especially long-term (Chawla, 1998; Palmer et al., 1999). This includes a teacher's choice of pedagogical approaches (Edwards, 2015). In a recent review of the literature by Hedefalk et al. (2014), three categories of the teacher's role from the teacher's perspective in environmental education were identified. These were (i) to educate children with facts *about* the environment, (ii) to influence and change children's behaviour, and iii) to encourage critical thinking.

More commonly, environmental education has been known through the three areas defined by the Lucas (1979) model of education *for*, *in* and *about* the environment (Edwards, 2015; Gough & Gough, 2010). Edwards (2015) highlighted Lucas' key argument that these three areas are "essential for holistic environmental education. This means that effective environmental education requires the deliberate inclusion and intent of education for the environment … as an integral component of all learning activities" (p. 16). In recent years, there has been increased interest and visibility towards education *in* the environment through the public promotion and campaigning by journalists such as Louv (2006) on the "new nature movement" which began through the work of the Scandinavian Forest Kindergartens and Schools and burgeoning research to this extent (e.g., see Blenkinsop et al., 2020; Kahn & Kellert, 2002; Malone, 2018; Malone et al., 2015; Sobel, 2008). It is worth noting that this work is problematic in perpetuating the human-centred model of relations with nature (Rautio, 2013b) along with limiting future possibilities through constantly focusing on the past (Malone, 2016).

Despite this movement, there are still many perceived barriers for teachers to educate *in* nature, such as the inability to control students, lack of resources, a crowded curriculum, changing educational aims, safety issues, lacking skills, lacking confidence, pedagogical knowledge and added paperwork (Barfod, 2018; Ernst & Tornabene, 2012; Glackin, 2018; McFarland & Laird, 2020; Simmons, 1998). Environmental education *for* and *about* the environment is comparatively less visible in recent research literature; however, past studies have identified that "perceived time constraints, over-crowded curriculum, constant change and lack of knowledge of environmental education as the major

barriers preventing or limiting the implementation of environmental education" (Cutter-Mackenzie & Smith, 2003, p. 519)
across all three domains along with lack of teachers' confidence
about their environmental knowledge (Gough & Gough, 2010).

Further to this, it was identified in a survey of approximately
5000 teachers in 2014 that 80 per cent of teachers do not comprehensively understand what education *for* sustainability is (Australian Education for Sustainability Alliance [AESA], 2014). In
the same year, Hill et al. (2014) conducted research with those
working with children from birth to eight years including teachers, educators, pre-service educators and parents to ascertain their
conceptualisation of education for sustainability. The findings revealed that the majority described that sustainability was about
the environment, in particular, garden-based activities (such as a
vegetable garden, a composting program, keeping chickens and a
worm farm), recycling and reusing. This study suggested that

> Finding innovative and transformative ways to embrace sustain-
> ability is one of the world's most pressing issues. We believe that
> sustainability education must start early and positively in the
> lives of young children, before unsustainable patterns of think-
> ing and acting are accepted as the everyday norm and become
> deeply ingrained habits – both children and the Earth deserve
> this commitment and action.
>
> Hill et al. (2014, p. 21)

The three categories of *for, in* and *about* were contentious at
the time they were proposed with concerns about their validity
(Edwards, 2015). It has been suggested that a fourth category,
"with" the environment also be included (Payne, 1998; Stevenson,
2007). Blenkinsop and Piersol (2013) suggest a "speaking with and

listening to" the more-than-human or nonhuman-other approach when considering environmental education pedagogical practices (p. 44). The original three-pronged approach has also been critiqued through a recent study in the emergent field of climate change educational research where education *for* sustainability or the environment was called a "moot point" for failing to address the complexities and possibilities associated with environmental education (Cutter-Mackenzie & Rousell, 2018, p. 12). More recently, and through the adoption of ethico-onto-epistemologies and the work of Baradian concepts being applied to environmental education research, the notion of education *as* sustainability is becoming evident in the research literature to promote the enactment of a relational ontology where human and nonhumans are materialised for transformative change (O'Neil, 2018).

In considering the implementation of current environmental education research into practice, the challenge for teachers has been to translate theory into practice – the theory-practice gap, otherwise known as the "rhetoric-reality" gap (Edwards, 2015; Grace & Sharp, 2000; Stevenson, 2007). Stevenson (2007) stressed that one of the key barriers to implementing environmental education into schools involves problematising the role of the teacher from "the dominant conception, organisation and transmission of knowledge" (p. 151). Edwards (2015) furthers this idea, by proclaiming that behaviourally changing teacher practice is challenging because of the "inability to readily embrace such change and to reimagine the purposes and practices of education is illustrated by the development of educational rhetoric–reality gaps" (p. 4). In addition, Robertson and Krugly-Smolska (1997) state that teachers are challenged by three perceived barriers: (i) practical issues, for example, time; (ii) conceptual issues, such as what environmental education *actually* entails; and (iii) whether they have permission to teach the often controversial environmental education subject. However, Robertson and Krugly-Smolska (1997) highlight that these barriers should be the work of researchers to mediate. That is, how are researchers framing their work so that it is (i) addressing the real and immediate concerns of teachers and teaching practice; (ii) making clear what environmental education is *actually* about; and (iii) providing teachers with explicit permission at the curriculum development level to do what researchers propose. However, Hart (2010) argues for a reconceptualisation of the "theory-practice gap" that adopts a relational epistemology where "knowing and learning involves an interplay of theories that guide action within the structures (institutions and cultural narratives) that surround such relations" (p. 165). The emphasis on the gap is not heavily placed on the deficit knowledge of the teacher but on a greater understanding of the social/relational processes that form the basis for education. In doing so, Hart (2010) is advocating for a dissolution of the binaries of theory/practice by proposing that there is another opportunity for understanding the relationship between these two entities. Hart (2010) continues that "if teacher education does not include elements of both critical reflection and social critique at several levels of engagement, then educational change is unlikely" (p. 165).

Conclusion

In considering ways to address the current global, environmental urgencies, this Turn has explored how early school years' teachers' perceptions of nature inform their pedagogy. I proposed that identifying the relationships and intra-activity between perception and pedagogy will further knowledge that seeks to understand if, and how, teachers know (they are) nature and what this means in practice. To ground this work, in this turn, I present/ed past research literature in environmental education and teachers' perceptions and pedagogies with/as nature.

Initially, the neoliberal, capitalist structure of minority education systems was problematised, and in doing so, I demonstrated that educational futures need to be aware of the limitations and inequalities of systems governed by this paradigm. Relationality through collaboration and togetherness was emphasised as a mechanism for mitigating the influence of capitalism. This focus on the relationality of human/nonhuman was again brought to the fore in the environmental education research that further identified learning and learning environments as a place for relationality to be explored.

Through analysing teachers' environmental education perceptions and pedagogies, it was identified that research in the field needs to look beyond the three-pronged approach to environmental education and further still beyond education *with* the environment to consider education *as* the environment. Through foregrounding the significance and importance of a relational approach to environmental education research and practice, the literature in this section highlighted the significance of unsilencing teachers' own knowing along with that of First Nations' knowledges. There is much work to be done here.

Notes

1 In this book, chapters are presented as a series of Turns. This is discussed in greater depth in the First Turn.
2 It is acknowledged that this terminology could be seen to promote binary thinking however is used thoughtfully through a posthuman lens and troubles the limitations of language through its use. The forward slash (/) is used to denote a sameness or no space or separation between the two terms.

References

Aguirre, J., & Speer, N. M. (1999). Examining the relationship between beliefs and goals in teacher practice. *The Journal of Mathematical Behavior, 18*(3), 327–356. https://doi.org/10.1016/S0732-3123(99)00034-6

Australian Education for Sustainability Alliance [AESA]. (2014). *Education for sustainability and the Australian Curriculum project: Final report for research phases 1–3.*

Barad, K. (2007). *Meeting the universe halfway: Quantum physics and the entanglement of matter and meaning.* Duke University Press.

Barfod, K. S. (2018). Maintaining mastery but feeling professionally isolated: Experienced teachers' perceptions of teaching outside the classroom. *Journal of Adventure Education and Outdoor Learning, 18*(3), 201–213.

Blenkinsop, S., Jickling, B., Morse, M., & Jensen, A. (2020). Wild pedagogies: Six touchstones for childhoodnature theory and practice. In A. Cutter-Mackenzie, K. Malone, & E. Barratt Hacking (Eds.), *Research handbook on childhoodnature: Assemblages of childhood and nature research* (pp. 1–18). Springer International Publishing.

Blenkinsop, S., & Piersol, L. (2013). Listening to the literal: Orientations towards how nature communicates. *Phenomenology & Practice, 7*(2), 41–60.

Blom, S. M. (2020). Conceptualizing parent(ing) childhoodnature through significant life experience. In A. Cutter-Mackenzie, K. Malone, & E. Barratt Hacking (Eds.), *Research handbook on childhoodnature: Assemblages of childhood and nature research* (pp. 1–26). Springer International Publishing. https://doi.org/10.1007/978-3-319-51949-4_127-1

Bookchin, M. (1987). Social ecology versus deep ecology: A challenge for the ecology movement. *Green perspectives: Newsletter of the green program project, 1987* (pp. 4–5).

Braidotti, R., & Dolphijn, R. (2017). *Philosophy after nature.* Rowman & Littlefield International.

Calderhead, J. (1996). Teachers: Beliefs and knowledge. In D. C. Berliner & R. C. Calfee (Eds.), *Handbook of educational psychology* (pp. 709–725). Prentice Hall International.

Carson, R. (1965). *Sense of wonder*. Harper & Row.

Chawla, L. (1998). Significant life experiences revisited: A review of research. *Journal of Environmental Education, 29*(3), 11. http://ezproxy.scu.edu.au/login?url=http://search.ebscohost.com/login.aspx?direct=true&db=pbh&AN=340880&site=ehost-live

Clark, C. M., & Peterson, P. L. (1984). Teachers' thought processes [Occasional Paper No. 72, Institute for Research on Teaching, Michigan State University].

Collingwood, R. G. (1960). *The idea of nature*. Oxford University Press.

Cronon, W. (1996). *Uncommon ground: Rethinking the human place in nature*. W. W. Norton & Company.

Cutter-Mackenzie, A., & Edwards, S. (2014). Everyday environmental education experiences: The role of content in early childhood education. *Australian Journal of Environmental Education, 30*(1), 127.

Cutter-Mackenzie, A., & Rousell, D. (2018). Education for what? Shaping the field of climate change education with children and young people as co-researchers. *Children's Geographies*, 1–15. https://doi.org/10.1080/14733285.2018.1467556

Cutter-Mackenzie, A., & Smith, R. (2003). Ecological literacy: The "missing paradigm" in environmental education (part one). *Environmental Education Research, 9*(4), 497–524.

Cutter-Mackenzie-Knowles, A., Brown, S. L., Osborn, M., Blom, S. M., Brown, A., & Wijesinghe, T. (2020). Staying-with the traces: Mapping-making posthuman and indigenist philosophy in environmental education research. *Australian Journal of Environmental Education, 36*(2), 105–128.

Davis, J. M., & Elliott, S. (2003). *Early childhood environmental education: Making it mainstream*. Early Childhood Australia, Inc.

Department of Education and Training [DET]. (2018). *Through growth to achievement: Report of the review to achieve educational excellence in Australian schools*. Commonwealth of Australia.

Desjean-Perrotta, B., Moseley, C., & Cantu, L. E. (2008). Preservice teachers' perceptions of the environment: Does ethnicity or dominant residential experience matter? *The Journal of Environmental Education, 39*(2), 21–32.

Devall, B., & Sessions, G. (1985). *Deep ecology: Living as if nature mattered*. Gibbs Smith.

Dewey, J. (1916). *Democracy and education: An introduction to the philosophy of education*. The Macmillan Company.

Dyment, J., Hill, A., & Emery, S. (2015). Sustainability as a cross-curricular priority in the Australian curriculum: A Tasmanian investigation. *Environmental Education Research, 21*(8), 1105–1126.

Edwards, J. (2015). *Socially-critical environmental education in primary classrooms: The dance of structure and agency*. Springer. https://doi.org/10.1080/09575146.2015.1064099

Edwards, S., Skouteris, H., Cutter-Mackenzie, A., Rutherford, L., O'Conner, M., Mantilla, A., Morris, H., & Elliot, S. (2015). Young children learning about well-being and environmental education in the early years: A funds of knowledge approach. *Early Years: An International Research Journal, 36*(1), 33–50. https://doi.org/10.1080/09575146.2015.1064099

Ernst, J. (2014). Early childhood educators' use of natural outdoor settings as learning environments: An exploratory study of beliefs, practices, and barriers. *Environmental Education Research, 20*(6), 735–752.

Ernst, J., & Tornabene, L. (2012). Preservice early childhood educators' perceptions of outdoor settings as learning environments. *Environmental Education Research, 18*(5), 643–664. https://doi.org/10.1080/13504622.2011.640749

Evans, N., Whitehouse, H., & Hickey, R. (2012). Pre-service teachers' conceptions of education for sustainability. *Australian Journal of Teacher Education (Online), 37*(7), 1–12.

Evernden, N. (1992). *The social creation of nature*. The JHU Press.

Fägerstam, E. (2012). Children and young people's experience of the natural world: Teachers' perceptions and observations. *Australian Journal of Environmental Education, 28*(1), 1–16. https://doi.org/10.1017/aee.2012.2

Flogaitis, E., & Agelidou, E. (2003). Kindergarten teachers' conceptions about nature and the environment. *Environmental Education Research, 9*(4), 461–478. https://doi.org/10.1080/1350462032000126113

Fraser, J., Gupta, R., & Krasny, M. E. (2015). Practitioners' perspectives on the purpose of environmental education. *Environmental Education Research, 21*(5), 777–800.

Froebel, F. (2005). *The education of man* (W. N. Hailmann, Trans.). Dover Publications.

García-González, E., Jiménez-Fontana, R., & Azcárate, P. (2020). Education for sustainability and the sustainable development goals: Pre-service teachers' perceptions and knowledge. *Sustainability, 12*(18), 7741.

Glackin, M. (2018). "Control must be maintained": Exploring teachers' pedagogical practice outside the classroom. *British Journal of Sociology of Education, 39*(1), 61–76.

Gough, N., & Gough, A. (2010). Environmental education. In C. Kridel (Ed.), *The Sage encyclopedia of curriculum studies* (Vol. 1, pp. 339–343). Sage.

Grace, M., & Sharp, J. (2000). Exploring the actual and potential rhetoric-reality gaps in environmental education and their implications for pre-service teacher training. *Environmental Education Research, 6*(4), 331–345.

Haraway, D. (2012). Awash in urine: DES and Premarin® in multispecies response-ability. *Women's Studies Quarterly, 40*(1/2), 301–316.

Haraway, D. (2015). Anthropocene, Capitalocene, Plantationocene, Chthulucene: Making kin. *Environmental Humanities, 6*(1), 159–165.

Hart, P. (2010). No longer a "little added frill": The transformative potential of environmental education for educational change. *Teacher Education Quarterly, 37*(4), 155–177.

Hatcher, A. (2012). Building cultural bridges with aboriginal learners and their "classmates" for transformative environmental education. *Journal of Environmental Studies and Sciences, 2*(4), 346–356.

Hedefalk, M., Almqvist, J., & Östman, L. (2014). Education for sustainable development in early childhood education: A review of the research literature. *Environmental Education Research*, 1–16. https://doi.org/10.1080/13504622.2014.971716

Hill, A., Nailon, D., Getenet, S., McCrea, N., Emery, S., Dyment, J., & Davis, J. M. (2014). Exploring how adults who work with young children conceptualise sustainability and describe their practice initiatives. *Australasian Journal of Early Childhood, 39*(3), 14–22.

Ingold, T. (2014). That's enough about ethnography! *Hau: Journal of Ethnographic Theory, 4*(1), 383–395.

IUCN. (1970). *International working meeting on environmental education in the school curriculum.*

Jickling, B., Blenkinsop, S., Morse, M., & Jensen, A. (2018). Wild pedagogies: Six initial touchstones for early childhood environmental educators. *Australian Journal of Environmental Education, 34*(2), 159–171. https://doi.org/10.1017/aee.2018.19

Kahn, P. H., & Kellert, S. R. (2002). *Children and nature: Psychological, sociocultural, and evolutionary* (P. H. Kahn & S. R. Kellert, Eds.). Massachusetts Institute of Technology.

Kimaryo, L. (2011). *Integrating environmental education in primary school education in Tanzania: Teachers' perceptions and teaching practices.* Åbo Akademi University Press.

Klees, S. J. (2019). Capitalisation and global education reform. In K. J. Saltman & A. J. Means (Eds.), *The Wiley handbook of global educational reform.* John Wiley & Sons.

Latour, B. (2004). *The politics of nature.* Harvard University Press.

Leopold, A. (1949). *A sand county almanac.* Oxford University Press.

Louv, R. (2006). *Last child in the woods: Saving our children from nature-deficit disorder.* Atlantic Books.

Lucas, A. (1979). *Environment and environmental education: Conceptual issues and curriculum implications.* College of Education, Ohio State University.

Malm, A., & Hornborg, A. (2014). The geology of mankind? A critique of the Anthropocene narrative. *The Anthropocene Review, 1*(1), 62–69. https://doi.org/10.1177/2053019613516291

Malone, K. (2016). Reconsidering Children's encounters with nature and place using posthumanism. *Australian Journal of Environmental Education, 32*(1), 1–15. https://doi.org/10.1017/aee.2015.48

Malone, K. (2018). *Children in the Anthropocene: Rethinking sustainability and child friendliness in cities.* Palgrave Macmillan, Springer Nature.

Malone, K., Birrell, C., Boyle, I., & Gray, T. (2015). Wild nature play: Researching out of school hours in the bush. *Sydney, Australia: Centre for Educational Research, University of Western Sydney.*

Maurer, M., & Bogner, F. X. (2020). First steps towards sustainability? University freshmen perceptions on nature versus environment. *PLoS ONE, 15*(6), e0234560. https://doi.org/10.1371/journal.pone.0234560

McFarland, L., & Laird, S. G. (2020). "She's only two": Parents and educators as gatekeepers of children's opportunities for nature-based risky play. *Research handbook on childhoodnature: Assemblages of childhood and nature research* (pp. 1075–1098). Springer International Publishing.

Montessori, M. (1917). *The advanced Montessori method* (Vol. 1). Frederick A. Stokes Company.

Moore, J. W. (2016). *Anthropocene or Capitalocene?: Nature, history, and the crisis of capitalism*. PM Press.

Moore, J. W. (2017). The Capitalocene, part I: On the nature and origins of our ecological crisis. *The Journal of Peasant Studies, 44*(3), 594–630.

Munoz, F., Bogner, F., Clement, P., & Carvalho, G. S. (2009). Teachers' conceptions of nature and environment in 16 countries. *Journal of Environmental Psychology, 29*(4), 407–413. https://doi.org/10.1016/j.jenvp.2009.05.007

O'Neil, J. K. (2018). Transformative sustainability learning within a material-discursive ontology. *Journal of Transformative Education, 16*(4), 365–387.

Orr, D. W. (1992). *Ecological literacy: Education and the transition to a postmodern world*. State University of New York Press.

Palmer, J. A. (2002). *Environmental education in the 21st century: Theory, practice, progress and promise*. Routledge.

Palmer, J., Suggate, J., Robottom, I., & Hart, P. (1999). Significant life experiences and formative influences on the development of adults' environmental awareness in the UK, Australia and Canada. *Environmental Education Research, 5*(2), 181–200. https://doi.org/10.1080/1350462990050205

Payne, P. (1998). Children's conceptions of nature. *Australian Journal of Environmental Education, 14*, 7.

Qiong, O. (2017). A brief introduction to perception. *Studies in Literature and Language, 15*(4), 18–28.

Quinn, F., Castéra, J., & Clément, P. (2016). Teachers' conceptions of the environment: Anthropocentrism, non-Anthropocentrism, anthropomorphism and the place of nature. *Environmental Education Research, 22*(6), 893–917. https://doi.org/10.1080/13504622.2015.1076767

Rautio, P. (2013a). Being nature: Interspecies articulation as a species-specific practice of relating to environment. *Environmental Education Research, 19*(4), 445–457. https://doi.org/10.1080/13504622.2012.700698

Rautio, P. (2013b). Children who carry stones in their pockets: On autotelic material practices in everyday life. *Children's Geographies, 11*(4), 394–408.

Robertson, C. L., & Krugly-Smolska, E. (1997). Gaps between advocated practices and teaching realities in environmental education. *Environmental Education Research, 3*(3), 311–326.

Rousell, D., & Cutter-Mackenzie-Knowles, A. (2020). Uncommon worlds: Toward an ecological aesthetics of childhood in the Anthropocene. In A. Cutter-Mackenzie-Knowles, K. Malone, & E. Barratt Hacking (Eds.), *Research handbook on childhoodnature: Assemblages of childhood and nature research* (pp. 1–23). Springer International Publishing. https://doi.org/10.1007/978-3-319-51949-4_88-2

Rousseau, J.-J. (1976). *Emile, or on education.* Basic Books. http://ww2.chandler.k12.az.us/cms/lib6/AZ01001175/Centricity/Domain/963/Rousseau%20Essay%20and%20Target%20Notes.pdf; https://books.google.ca/books?id=VocWKgK9SxQC&printsec=frontcover&source=gbs_ge_summary_r&cad=0#v=onepage&q&f=false; http://lf-oll.s3.amazonaws.com/titles/2256/Rousseau_1499_Bk.pdf

Sheldrake, R. (1990). *The rebirth of nature.* Random Century Group Ltd.

Simmons, D. (1998). Using natural settings for environmental education: Perceived benefits and barriers. *The Journal of Environmental Education, 29*(3), 23–31. https://doi.org/10.1080/00958969809599115

Smith, K., Tesar, M., & Myers, C. Y. (2016). Edu-capitalism and the governing of early childhood education and care in Australia, New Zealand and the United States. *Global Studies of Childhood, 6*(1), 123–135. https://doi.org/10.1177/2043610615625165

Sobel, D. (2008). *Childhood and nature: Design principles for educators.* Sternhouse Publishers.

Soper, K. (1995). *What is nature?* Blackwell Publishers.

Stanford University. (2022). *Stanford encyclopedia of philosophy* (E. N. Zalta, Ed.). The Metaphysic Research Lab, Philosophy Department, Stanford University.

Stehlik, T. (2018). *Educational philosophy for 21st century teachers.* Palgrave Macmillan.

Stevenson, R. B. (2007). Schooling and environmental education: Contradictions in purpose and practice. *Environmental Education Research, 13*(2), 139–153.

Stevenson, R. B., Ferreira, J.-A., & Emery, S. (2016). Environmental and sustainability education research, past and future: Three perspectives from late, mid, and early career researchers. *Australian Journal of Environmental Education, 32*(1), 1–10.

Taylor, A. (2013). *Reconfiguring the natures of childhood.* Routledge.

Taylor, A. (2017). Beyond stewardship: Common world pedagogies for the Anthropocene. *Environmental Education Research, 23*(10), 1448–1461. https://doi.org/10.1080/13504622.2017.1325452

Thi To Khuyen, N., Van Bien, N., Lin, P.-L., Lin, J., & Chang, C.-Y. (2020). Measuring Teachers' perceptions to sustain STEM education development. *Sustainability, 12*(4), 1531. https://www.mdpi.com/2071-1050/12/4/1531

Thoreau, H. D. (2004). *Walden: A fully annotated edition.* Yale University Press.

Torquati, J., Cutler, K., Gilkerson, D., & Sarver, S. (2013). Early childhood Educators' perceptions of nature, science, and environmental education. *Early Education and Development, 24*(5), 721–743. https://doi.org/10.1080/10409289.2012.725383

UNESCO. (1976). The Belgrade charter. *Connect: UNESCO, UNEP Environmental Education Newsletter* (pp. 1–2).

Wallerstein, I. M. (2004). *World-systems analysis: An introduction.* Duke University Press.

Waters-Adams, S. (2006). The relationship between understanding of the nature of science and practice: The influence of teachers' beliefs about education, teaching and learning. *International Journal of Science Education, 28*(8), 919–944. https://doi.org/10.1080/09500690500498351

Whitehouse, H., Watkin Lui, F., Sellwood, J., Barrett, M., & Chigeza, P. (2014). Sea country: Navigating indigenous and colonial ontologies in Australian environmental education. *Environmental Education Research, 20*(1), 56–69.

3

THE THIRD TURN

Posthuman theorising in educational research with teachers' nature perceptions and pedagogies

Introduction

Doing theory requires being open to the world's aliveness, allowing oneself to be lured by curiosity, surprise, and wonder. Theories are not mere metaphysical pronouncements on the world from some presumed position of exteriority. Theories are living and breathing reconfigurings of the world. The world theorizes as well as experiments with itself. Figuring, reconfiguring.

Barad (2012, p. 2)

In this Turn, as we continue to explore teachers' perceptions of nature and how this informs their pedagogy, we turn to theory and look to posthuman thinking. As is detailed in this Turn, posthuman theory is conceptualised in many ways. The posthuman theoretical perspective underpinning this book was inspired by the work of quantum physicist and feminist theorist, Karen Barad (2007). Barad's (2007) theorising provides a fierce and rigorous underpinning for challenging human-centrism while

DOI: 10.4324/9781032703473-3

acknowledging that "the human is always already implicated in knowledge-making practices" (Murris, 2022, p. 1). As Barad (2012) states in the opening vignette, theories are alive, dynamic, and invite all the human/nonhuman world players to be involved. Barad (2007) adopts this position in proposing their rich and complex theories which are ethico-onto-epistemologically situated. These theories have much to offer educational research, as Murris (2022) highlights, Barad's work proposes "alternatives to the dominant neoliberal national curricula, educational policies, and humanist teaching, research and conference agendas" (p. ix). Barad's (2007) theories which problematise human centredness and traditional educational agendas, provide a fitting foundation in posthuman research. As such, I draw specifically upon Barad's concepts of material-discursive practices and ethico-onto-epistemology to inform the theoretical position for this work.

A note on ethico-onto-epistemology

Ethico-onto-epistemology was first proposed by Barad (2007) as a way to decentre the human and more aptly describe how practices of knowing "cannot fully be claimed as human practices … because knowing is a matter of part of the world making itself intelligible to another part" (p. 185). From a quantum physics perspective, this is about the communication, the intra-actions that are constantly occurring at the micro and macro levels between particles, irrespective of whether those particles are human or nonhuman. That is, ways of knowing and being in the world are not restricted to the human form. "Practices of knowing and being are not isolable; they are mutually implicated" (Barad, 2007, p. 185), they involve practices that are omnipresent for human/

nonhuman alike, and they are always already entangled in/with/ through each other.

The main purpose of an onto-epistemology is to trouble dichotomies, such as the conventional views of ontology and epistemology that consider ways of knowing the world and being in the world to be separate schools of thought. However, "we don't obtain knowledge by standing outside the world; we know because we are of the world. We are part of the world in its differential becoming" (Barad, 2007, p. 185). To authentically embrace research that challenges binaries, including those that separate reality and information, "in other words between ontology and epistemology" (Barad, 2007, p. 32), it is necessary to adopt an onto-epistemology. Zeilinger (2010) also argued for disturbing dualisms through an onto-epistemological stance, "that is neither purely ontological nor purely epistemological" (p. 38).

Tracings of the philosophies and quantum theories proposed by Barad have been documented by many philosophers, scientists, and theorists throughout history, however, what distinguishes Barad's ideas apart is that they are grounded in "rigorous accounts of the experimental practice of physics" (Snaza et al., 2016, p. xvii), they are meaningful in the context of spacetime, and they give attention to the detail and minutiae of words and life. Adopting an ethico-onto-epistemology for inquiring into pedagogical practices is innovative in pushing back against the conventional practices of separate ontologies and epistemologies in educational research; however, it is not without challenges. Pushing up and back against traditional approaches requires a willingness to venture into the "unknowns". This includes an education that Snaza and Weaver (2014) argue would do well to not only consider human bodies but the "ecological networks that include humans, animals, an enormous number of inanimate objects, and the myriad political relations obtaining among all of these" (p. 27). This draws on the ethical framework that contributes to an ethico-onto-epistemology where relationality is an intrinsic part of the learning and, of the becoming. The concepts proposed through an ethico-onto-epistemology enable the data in this research – namely, the lesson participations, the video-stimulated recalled conversations and visual journal entries – to be viewed from the perspective of the agency of matter, thus allowing a disrupting of human privileging and forming an appropriate framework for exploring early school years teachers' perceptions of nature and how they inform pedagogy. Barad (2007) describes this framework succinctly and almost poetically,

> …what we need is something like an ethico-onto-epistem-ology – an appreciation of the intertwining of ethics, knowing, and being – since each intra-action matters, since the possibilities for what the world may become call out in the pause that precedes each breath before a moment comes into being and the world is remade again, because the becoming of the world is a deeply ethical matter.
>
> Barad (2007, p. 185)

From this ethico-onto-epistemological viewpoint, research is not conducted from an exteriority or privileged position of the "all-knowing researcher", but as a humble researcher participant who is open, alive, and receptive to the intra-actions that research practices enact. This framework became my ethics of responsibility as a researcher participant – my response-ability – to listen for the "call out in the pause" (Barad, 2007, p. 185), and respond.

In this Turn, posthuman theory is presented as it informs the overarching framing of the work this book is founded upon. The posthuman framework is then detailed to demonstrate the series of three theoretical offerings that are informed by the concepts of material-discursive practices, affective atmospheres, and childhoodnature. These three concepts constituted the Returning Learning theoretical framework which informed the diffraction gratings to make the diffractive data entanglements – the intraconnected data collection and analysis processes. These three concepts were suitably applied to this study as they challenge dominant humanist approaches to education, problematise binary-making practices, and resist conventional research paradigms; such tenets are considered necessary and urgent in enacting response-abilities to the current global catastrophe.

Posthumanism

Posthuman theory emerged at the end of the 19th century as a remedy to the domination of humanist theories on educational research and practice. This relatively and so-called new theoretical perspective is primarily concerned with the deconstruction of binary thinking and, as the name suggests, moving beyond human privileging and human-centredness (Castree & Nash, 2006; Jansen et al., 2021; Miah, 2008; Snaza & Weaver, 2014). However, I trouble the term "new" I have used here in describing posthuman theory. Indeed, Barad (2014) argues that "there is nothing that is new; there is nothing that is not new" as there is "no absolute boundary between here-now and there-then" (p. 168). While I dwell in the tension of the not-new "new" and emphasise the word "new" throughout this book, I acknowledge that many of the tenets of posthumanism are eons old and should be attributed to their rightful lineages. "All entities have their antecedents; they emerge from things that came before them" (Prout, 2011, p. 4) and posthumanism indeed has traces throughout history. I begin with First Nations Peoples, cultures, and knowledges. This includes concepts of relationality, interconnectedness and living in a way such that there was no exterior, no external environment – nonhuman nature needed no distinct referent as it was considered an entangled part of the whole (Blaikie et al., 2020; Cutter-Mackenzie-Knowles et al., 2020a; Murris, 2020b). Moreover, the posthuman work in troubling and problematising humanism for its tendencies to minoritise certain groups of people, is attributed to those, such as First Nations Peoples, that have engaged in these discourses for thousands of years (Snaza et al., 2014). Posthuman thinking in educational research is taking the ethical turn in respectfully acknowledging the First Nations' knowledges that precede it, for example, posthuman thinking acknowledging Canadian First Nations Peoples (Blaikie et al., 2020), African First Nations Peoples (Murris, 2020b), and Australian First Nations Peoples (Cutter-Mackenzie-Knowles et al., 2020a).

There are further antecedents to posthuman theory which are also acknowledged, as for posthumanism to exist, there first had to be the theory of humanism. Humanism, in Western minority world thinking, was introduced as a concept in the nineteenth century (Snaza et al., 2014). It was built on Ancient Greek philosophy that proposed a humanity where the human was not only biological *as an animal* but also social and political amongst other philosophical positionings that were contextualised in the historical, political, cultural, economic and social constructs of the time (Snaza et al., 2014). Humanism, then, was rife with conceptualisations of those who were not considered as human including women, black slaves, and colonised natives and as such, became regarded as "things", "objects to be used *by* humans" (Snaza et al., 2014, p. 42; emphasis added by author). This "epistemic violence" (Braidotti, 2013, p. 30) is precisely what posthumanism is driven to move beyond. However, this critique is again, not new nor is it attributed to posthuman thinking. Theorists from feminist, post-structuralist, postcolonial and queer theory have all challenged traditional humanist notions for their "exclusionary work done by adhering to a singular model of human identity" (Lorimer, 2009, p. 345). Indeed, the term "posthumanism" was coined in 1977 as a "broad speculative concept within postmodernism" (Jansen et al., 2021, p. 217).

Humanist thinking has been the dominant paradigm in educational theorising since its inception (Snaza, 2014), and while posthumanism in Western minority world thinking has only risen in visibility over the last three decades, the notion of questioning the rational, "human-centric" model of the Enlightenment period began with the Romantics (Castree et al., 2004). Castree et al. (2004) emphasise how the term "post-Human" was presented in 1888 by Helena Blavatsky, and as such, and again, there is nothing *new* about posthumanism, even though these early postulations held a different focus to the "scholarly culture of post-war posts" (Jansen et al., 2021, p. 231) such as posthumanism that is the focus in this book.

Snaza et al. (2014) highlight the challenge in defining posthumanism given its multi-disciplinary and historical inputs, and offer a series of propositions to conceptualise the posthuman including as a way to challenge humanist politics, as an entanglement of human and technologies, and through animal and curriculum studies. Miah (2008) purports that posthumanism has two trains of thought: one as cultural which is concerned with destabilising human-centred values, and the other as philosophical, which incorporates the idea of the human-technological hybrid, or the "transhuman" (such as the work in referent to cyborgs – for instance, Donna Haraway's Cyborg Manifesto – attends to). Jansen et al. (2021) and Petitfils (2014) similarly attest to this two-fold way of interpreting current posthumanist thinking. Lorimer (2009) categorises posthumanism into a series of four modalities across a section of five posthuman themes. However, this divisive categorisation seems to move against the cohesive intra-connectedness that posthumanist thinking argues for.

Posthuman theory has risen in the last 40 years from human geography and the social sciences to become more prevalent in educational research over the last decade through the pioneering work of Snaza and Weaver (2014), Snaza et al. (2014), and Murris (2016). These researchers have drawn on many lines of thinking, most prominently the works of Haraway (1992), Braidotti (2013), and Barad (2007). In this book, I have picked up the thread from these posthuman thinkers and a posthuman position that advocates for and proposes an educational scape that challenges the human – or as Miah (2008) states – a cultural posthumanism.

As Barad (2007) purports, posthuman theory "is about taking issue with human exceptionalism while being accountable for the role we play in the differential constitution and differential positioning of the human among other creatures (both living and nonliving)" (p. 136). As with many concepts presented by Barad (2007), posthuman theory through Barad's work, is underpinned by response-ability. In this case, an important and intrinsic acknowledgement of the human body's position in equanimity with the nonhuman. Snaza and Weaver (2014) argue that education's current saturation in humanism makes it near impossible to conceptualise a posthuman education – where a response-ability to educating from a theoretical position that embodies the equality of human/nonhuman is not a reality. This is the tension that posthumanism imposes, that, human bodies cannot ever not be human so can only propose theoretical approaches to dwell, reside, and "stay-with" (as described by Haraway, 2016) the tension. In conceptualising what this means for educational research and practice, Snaza and Weaver (2014, p. 9) offer that,

> ...whenever [educational] research is conducted in schools there is much more going on than the interaction between a teacher and students or teachers and teachers or students and students. There are experiences happening all the time, all over the school, independent of humans. There are always interactions between humans and nonhuman sentient beings and humans and nonsentient objects, such as computers, doors, playgrounds, hallways, utensils, trays, balls, windows, desks, and so on.

Taylor (2013) similarly describes the agency and intra-activity of a teacher's chair, pen, and t-shirt in the classroom setting to "illuminate how that which is resolutely mundane within everyday pedagogic practice nevertheless possesses a surprising material force" (p. 689). Educational research that adopts this posthuman thinking that pushes beyond the boundaries of human superiority is (still) largely minoritised given the "enormous, almost crushing weight of several millennia of humanist thought" (Snaza & Weaver, 2014, p. 1). However, such research is desperately needed to think beyond the human form, while residing deep within it, to enact the possibilities of a relational future. A future where not only the seen material objects – human/nonhuman – are accounted for, but one where the unseen forces that inherently become material are equally considered. More on this is upcoming in the discussion on material-discursive practices.

Posthumanism is not polarised from humanism. On the contrary, it is an onward and outward progression. Some interpretations of posthumanism appear to wholly reject certain philosophical ideas attesting that they maintain individual identity and as such, perpetuate humanism (see Merrell, 2003). In accordance, Braidotti (2013) professes the overarching dominant paradigms of the minority-world social, cultural, and pedagogical ideals that infiltrate current practices being attributed to humanism; she also notes that humanism has some worthy fundamental principles, namely "responsibility, self-determination, solidarity, community-bonding, social justice and principles of equality" (p. 29). Braidotti (2013) further describes the value of humanist pragmatism through an "adventurous element, a curiosity-driven yearning for discovery and a project-oriented approach" (p. 29). Thrift (2008) concurs that discarding humanism entirely is problematic and attests to holding on to a degree of humanism, "no matter how faint" (p. 13). This practice of reconciling humanist and posthumanist oppositional tensions works with posthumanist theorising in grappling with binaries. Braidotti (2013) describes this as the core of posthumanism: the ability "to disengage the positive elements of Humanism from their problematic counterparts" (p. 30). This thinking is confirmed by Murris (2016, p. 46) who attests,

> The "post" in "posthumanism" does not suggest that we can or should leave "humanism" behind – even if this were possible. The past is not fixed (like static events "in" time and space without always being part of the present and the future). Humanism refers to "something essential and universal, with a defining quality" that is shared by all human animals, and identity is a "humanist signifier".
>
> Jackson & Mazzei (2012, p. 68)

Posthumanism is a process. It is a continual unfoldment of redefining, reconceptualising, and reconfiguring. This process is described in many posthuman accounts as the (re)turn – the turning over and over – the practice of continual discarding and acquiring (for example, see Barad, 2014; Malone, 2018). Snaza (2013) describes this as a process of learning "how to 'let go' of being 'human'" (p. 51). Indeed, Snaza (2013) proposes education's reconceptualisation as a process that leads humans away from "the stable, predictable, and cultured world of civilization, of cities, of routine, of politics as we have known it. Whither it should lead us is—and must be—unknown" (p. 11). This is at the very crux of posthuman theory – destabilising the core foundations of what humans have come to "know" such as through problematising the binaries that have defined the human against the nonhuman. The process of deconstructing dualisms is foundational to posthumanism and has been noted by many researchers (see, for example, Alaimo, 2010; Barad, 2007; Cutter-Mackenzie-Knowles et al., 2020a; Latour, 2004; Lenz Taguchi, 2010; Malone, 2018; Murris, 2020b; Tuana, 2008).

Murris and Borcherds (2019) describe a posthuman ontology as one that "disrupts the Nature/Culture binary on which modern schooling has been built" (p. 60) and maintain that any posthuman theorising should hold the intention to not perpetuate dualisms. Posthuman thinking, through Barad's (2007) quantum physics lens, seeks to diminish such distinctions, which aligns with the ethico-onto-epistemological positioning that refutes hierarchical prioritising.

Posthuman theory provides an opportune platform for deconstructing individualised disciplinary approaches to research and thus, promotes transdisciplinary mechanisms, or even antidisciplinary educational research and practice (Snaza & Weaver, 2014). In addition, posthumanism is also an enabler of traditionally excluded or labelled categories such as race, gender, class, ability, and age (Murris & Haynes, 2018b). As Murris and Haynes (2018b) describe, "posthumanism opens up fresh opportunities for doing research in teaching and learning" and "radically changes what it means to teach and what it means to learn" (p. 4). Ways of teaching and learning are complex phenomena and posthumanism offers a radical paradigm shift to explore the possibilities for learning and knowing through global, and arguably universal, perspectives in local contexts (Murris & Haynes, 2018b). Murris (2018) attests that "posthuman research requires an un/learning ... an unsettling of agency, voice and identity as not something subjects 'have'" (pp. 29–30). Blaise (2013) supports this view by discussing the recent shift by some educators and researchers from the dominant developmental conversations in education towards posthuman theories in offering new ways of considering teaching and learning. While megaliths such as the *Research Handbook on Childhoodnature*

(Cutter-Mackenzie-Knowles et al., 2020b) demonstrate the movement of educational research and practice towards posthuman thinking, publications such as the document prepared for the Canadian Commission for UNESCO (Blaikie et al., 2020), offer an example of the practical application of the potential of posthuman for teaching and learning practice for the future of education.

Building on previous works utilising posthuman thinking in environmental education (Änggård, 2016; Braidotti & Dolphijn, 2017; Crinall, 2017; Cutter-Mackenzie-Knowles et al., 2020b; Malone, 2018; Murris, 2020b; Rautio, 2013b; Verlie, 2018), this book provides further empirical research and greater contextualisation and uptake of these critically important theories for practice; not only the educational, but the ecological future depends upon it. In accepting this position of posthumanism, the inherent struggles and tensions are acknowledged and a kind of "staying-with" (Haraway, 2016) was adopted where I, in human form, cannot escape the humanising of my perception. However, through my awareness and willingness to confront and trouble this tension, I have asserted the possibilities and potentialities that ensue. As Murris (2020b, p. 50) attests to work that decentres the human and adopts diffractive pedagogies and research methodologies,

This is important work, as they show how a posthuman inclusion of the more-than-human can work in the context of schooling – rendering children "capable by and with both things and living beings" (Haraway, 2016, p. 16). The details of these material-discursive entanglements matter, and tracing them "link actual beings to actual response-abilities"

(Haraway, 2016, p. 29). When child is reconfigured … more equitable relationships between humans (of, e.g., different ages) and between the human and more-than-human are brought into existence.

Taking a conceptual turn: proposing the Returning Learning theoretical framework with material-discursive practices, affective atmospheres, and childhoodnature

As described previously in this Turn, three posthuman concepts underpin the Returning Learning theoretical position that frames this study. Firstly, material-discursive practices describe how *matter extends beyond the physical and has agency that communicates*. Secondly, affective atmospheres consider how *physical spaces hold a quality that can be felt*. Finally, childhoodnature explores how *binary-making practices are exposed, and troubled*.

Material-discursive practices: matter extends beyond the physical and has agency that communicates

Materiality describes objects through their matter rather than focusing on what that matter is. All matter matters and material-discursive practice introduces the idea that matter has agency – it builds on the idea that matter "speaks" – it is vibrant, lively, alive – and as stated by Barad (2007), has an "active factor in further materialisations" (p. 66). The activity of matter is termed material-discursive practices, or material-discursive forces. Material-discursive practices primarily explain how matter makes itself felt – the social, historical, cultural, economic, natural, physical, and geopolitical forces, amongst others – that are important in understanding the entangled nature and process of materiality (Barad, 2007). Being aware of and understanding the material-discursive practices that operate in the classroom is imperative to the work of a teacher in more deeply being able to respond to their students.

The process of understanding and (re)conceptualising discursive practices as extending to include materiality requires an understanding that discourse is not just "what is said" (Barad, 2007, p. 146) nor are they "speech acts, linguistic representations, or even linguistic performances, bearing some unspecified relationship to material practices" (Barad, 2007, p. 166). Reconceptualising notions of discourse involving only human conversations and representations is the initial step in reconfiguring an understanding and interpretation of the word "discourse". The ideas of discourse just being human perpetuate humanism and do not fit the conceptual frame where all matter is entangled through the processes of material-discursive practices.

In a similar context, Murris (2016) utilises the concept of material-discursive forces through her work in early childhood to coin the phrase "the posthuman child". Through this conceptualisation, Murris (2016) espouses that children are discursive, material, relational and that there is "equality between species and between different members of each species" (p. 193). As such, material-discursive forces work to challenge binaries and dualisms by promoting all matter – both human and nonhuman – as agentic. As such, even knowledge, which most humans hold so dear, cannot be owned or associated solely with the human object. Knowledge is co-constituted through material-discursive practices of humans and nonhumans alike (Taylor, 2016). In addition, O'Neil (2018) purports that "humans and nonhumans co-constitute new meaning through the iterative material-discursive learning process resulting in a transformation. In other words, human and nonhuman are agentic forces and agency emerges from this highly relational learning process" (p. 376). Through the embodiment and enactment of this conceptualisation of knowledge production as a co-generative process, material-discursive practices help to "unpick (some of) the damages done by colonialist, racist, humanist knowledge practices" and "promise to 'do knowledge differently' [by] engaging with the geopolitical materialisation … as an ethical and political imperative" (Taylor, 2020, p. 27).

As Taylor et al. (2020) propose, material-discursive practices promote research that aligns with the posthuman and "shifts the ontological locus of agency from the human and redistributes it within human-nonhuman" (p. 171) entanglements of matter and materiality. In enacting research that embodies this theorising, a different kind of response-ability is called for where participants are considered "beyond the human to include the nonhuman, things and nature" (Taylor et al., 2020, p. 171).

In earlier work, Murris et al. (2018) propose that "agency is not located in a person, and instead, is attributed to a complex field of forces" (p. 67), namely, material-discursive forces. More recently, Murris (2020a) expands on this idea by proposing that "the human is neither a biological nor an ontological given but a political concept and a material-discursive *doing*, not a *thing*" (p. 69, emphasis in original). Here, Murris (2020a) emphasises the materiality of bodies as practices (of *doings*) rather than the traditional focus on objects or mass. As is a tenet of quantum theory, it then makes sense that "darkness is not mere absence, but rather an abundance" (Barad, 2014, p. 171) where even space – which is so frequently considered as an inert background – is agentic with its own material-discursive practices (Menning et al., 2020). As such, material-discursive practices are "seen" to extend beyond the physical materiality of objects and exist regardless of spacetime.

Affective atmospheres: physical spaces hold a quality that can be felt

...affect is itself a real condition, an intrinsic variable of the late-capitalist system, infrastructural as a factory. Actually, it is beyond infrastructural, it is everywhere, in effect. Its ability to come second-hand, to switch domains and produce effects across them all, gives it a meta-factorial ubiquity. It is beyond infrastructural. It is transversal.

Massumi (1995, pp. 106–107)

Here, Massumi (1995) proposes the ubiquitous nature of affect, where affect can be used to explain the "invisible, nonrepresentational … part of the ubiquitous backdrop of everyday life" (Bissell, 2010, p. 272). Massumi (1995) further contributes that affect equates to intensity, and *not* emotion as it is frequently described. Emotion as described by Massumi (1995) is attributed to being subjective, and personal. However, and perhaps in avoiding creating yet another binary, Massumi (1995) does not propose that effect is objective. Instead, affect is described as "unqualifiable … not ownable or recognizable" (Massumi, 1995, p. 88). Zembylas (2021) highlights that effect "is a much broader term denoting modes of influence, movement, intensity, and change" (para. 2). Moreover, affect offers a "wider perspective on 'affect' highlighting difference, process, and force" (Zembylas, 2021, para. 2). This is explained by Dernikos et al. (2020a p. 5; emphasis in original):

...affects are the forces (intensities, energies, flows, etc.) that register on/with-in/across bodies to produce and shape personal/emotional experiences. In other words, affect is not what you feel, as much as it is an event that forces you to *be(come) affected*, to feel *some-thing*.

To "feel *some-thing*" is the way affects and feelings are sensed, registered and understood by the human body, and these embodied ways of knowing engage with "sensation, memory, perception, attention and listening" (Blackman & Venn, 2010, p. 8).

When affect is considered as part of the "affective turn", that has reinvigorated the social sciences over the last two decades, and education research over the recent few years, Zembylas (2021, para. 2) describes how,

> ...the affective turn in education expands our thinking and research by attempting to enrich our understanding of how teachers and students are moved, what inspires or pains them, how feelings and memories play into teaching and learning.... What the affective turn contributes to education and other disciplines is that it draws attention to the entanglement of affects and emotions with everyday life in new ways. More importantly, the affective turn creates important ethical, political, and pedagogical openings in educators' efforts to make transformative interventions in educational spaces.

The concept of "affective atmospheres" is derived from cultural studies (specifically, geography) to explain the "emergence and circulation of affects, which are embodied forces or intensities that are both produced by and produce bodies" (Verlie, 2018, p. 15). Affective atmospheres bring attention to the collective, non-representational background of "affective, embodied conditions for representational acts and practices" (Ash & Anderson, 2015, p. 34), and provide a theoretical backing to consider the "diffuse, collective nature of affective life" (Ash & Anderson, 2015, p. 34). Studies in affective atmospheres demonstrate the vast potential for understanding and articulating learning situations, and as such, this thinking is gaining momentum in educational research (see for example Dernikos et al., 2020b; Finn, 2016; Murris & Haynes, 2018a; Snaza, 2020; Verlie, 2018; Verlie & Blom, 2021; Wolfe & Rasmussen, 2020; Zembylas, 2020). This includes reframing the traditional humanist perspectives of school and classroom climate through the affective to consider the nonhuman influence (Mayes et al., 2020; Verlie & Blom, 2021).

Murris et al. (2018, p. 162), during their research in a South African, grade-two classroom utilised posthuman thinking and practice, including affective atmospheres. They note that,

> An atmosphere may feel static and sedimented, but they are always "on the move" and intra-acting with everything and "nothing".... We have come up with various contradictory ideas to name the atmospheres we sensed as we watched and diffracted with the footages: discomfort, awkwardness, surprise, embarrassment, horror, shame, amusement, belittlement, tension, conflict.

Murris et al. (2018) highlight that the atmospheres were "sensed" and the problematic nature of naming an atmosphere that is dynamic and always already reconfigured. I acknowledge the tensions that are created in applying a non-representational field of study where the products are intentionally named through the use of affective atmospheres (Murris et al., 2018). Similar to the tensions that are experienced in adopting a posthuman perspective from a human body, I dwell in and "stay-with" these tensions in these theoretical offerings.

Affective atmospheres are suitably nestled in the arms of posthuman theory as they similarly aim at decentring the individual as they "instead prompt us to think about how different configurations of objects, technologies, and bodies come together to form different experiences of 'being with'" (Bissell, 2010, p. 272). Here, Bissell (2010) highlights the importance of considering the various forms of matter that constitute and reconfigure spaces and the intra-action between these bodies. In addition, Barad (2012) explores the "material-affective dimensions" of touch to describe its "physicality, its virtuality, its affectivity, its e-motion-ality, whereby all pretence of being able to separate the affective from the scientific dimensions of touching falls away" (p. 3). Affective touching is not about physical objects coming together but understanding the fullness of space – as an abundance and not as a darkness (Barad, 2014). As such, reconfiguring space as more than just an inert background is important, as the affective atmospheres that constitute spaces are in constant communication, informing bodies and their movements in everyday life, including the classroom milieu. Affective atmospheres give voice to the collective of material-discursive practices and provide a mechanism to articulate the language of the differences that are occupied in affective spaces; for example, the classroom which is disproportionately focused on the human presence. This is why, as a concept, affective atmospheres has been fittingly put to work with material-discursive practices in the theoretical framework proposed in this book. From an affective atmospheres and material-discursive position, the humanistic tendencies of research traditions are decentred to include the agency of human/nonhuman bodies equally.

The final consideration in the theoretical entanglement is childhoodnature. Childhoodnature is an emergent concept that describes the inherent connection between childhood and nature from a posthuman frame. As such, it is suitably used alongside material-discursive practices and affective atmospheres.

Childhoodnature: binary-making practices are exposed, and troubled

Childhoodnature is a posthuman term that seeks to fracture boundaries between childhood and nature, and in this sense, childhoodnature is a collective. It was coined by contributors to the International Colloquium on Childhoodnature in 2015 (Cutter-Mackenzie-Knowles, Logan, Osborn, and Blom) that inspired the congregation of the pivotal *Research Handbook on Childhoodnature* edited by Cutter-Mackenzie-Knowles et al. (2020). This emergent field was brought to the fore to "respond effectively to what is regarded as rapidly changing conditions of life on earth for all species and things" (p. 2). Childhoodnature offers "theories which children themselves can use to address the crises which they will inevitably inherit (and already are)" (Malone et al., 2020, p. 2).

Childhoodnature theories allow an openness in exploring "what it means to be human, not as outside of the world but deeply entangled with all that makes up the human and more than human world" (Malone et al., 2020, p. 3). At the core, childhoodnature contests the child/nature binary (which could be considered more broadly through the human/culture or nature/culture divides) in response to the "new nature movement", and provides opportunity in the educational space for practices and pedagogies that disrupt, for example, human exceptionalism and age-discrimination (Malone et al., 2020; Murris, 2020b).

Through childhoodnature, "learning is understood in terms of different matter – human and nonhuman – making themselves intelligible to each other" (Somerville, 2020, p. 107).

Rautio (2013b) contributes to a childhoodnature onto-epistemology when she attests that children should be appreciated "as whole beings and not just waiting to become something" (p. 396). The "waiting-to-become-something" narrative is proliferated through the notion of the child as *becoming* that is suggested in many educational paradigms, especially in early childhood, that only look to the development of the body without considering the *being* within. This idea is not new however, and was brought to attention by the work of Dewey (1916) who reconceptualised childhood as "one that does not measure it according to a teleological narrative about becoming adult, but instead sees it as plenitude of pure potentiality" (Snaza, 2017, p. 28).

Rautio (2013b) elaborates on childhood agency by considering how it is used in understanding childhood such that children are "autonomous and independent in the sense that their material surroundings, human and nonhuman, would not play any part in the kind of beings they are" (p. 396). This materialist approach to understanding children and as such, childhood, is proposed to challenge the epistemic notions of agency so that it allocated "space in between children and their environments, arising in complex encounters rather than located only in the human individuals" (Rautio, 2013b, p. 396).

Murris (2020b) advocates for equitable research and practice in education highlighting how posthumanist theories such as childhoodnature support emergent and varied methodologies. Murris (2020b) drew on examples of this, however, the silence of empirical research in the early years of schooling was evident.

As suggested by Malone et al. (2020), past ecological and environmental theories have promoted a degree of challenging dominant paradigms in childhood research and practice through acknowledging the nature/culture and more specifically, the child/nature divide. These propositions challenge claims of "nature-deficit disorder" as proposed by Louv (2006) while also acknowledging the significance of the awe, wonder and majesty that nonhuman nature provides (Carson, 1965). Empirical data from Kalvaitis and Monhardt (2015) confirms that "participating children did not see themselves as separate from nature"; however, they also describe the significance of nonhuman nature experiences through one of their participants' responses: "'I don't spend every second outside, but when I am outside I almost feel like, like I'm alone with nature, like I become nature' (fifth-grade child)" (para. 41). Human physicality in nonhuman nature spaces provides an attunement, a reminder of human natureness, of childhoodnature. Bai et al. (2010) explore the notion that childhoodnature through the idea of nature being educated out of and away from children, and dispute that education should be founded upon children bringing attention to themselves through "being sense, being bodies, being perceptions, being feelings" (p. 36). Research by Loughland et al. (2002) supports this idea demonstrating that young children are more relational and less object-focused than their older counterparts.

In a posthuman theoretical turn, Snaza (2017) describes how Dewey and Bennett were also aware of the humanist ideals of education and that "learning to come to terms with a more-than-human democracy requires re-thinking and re-valuing the 'experience' of children before they are educated into humanist anthropocentrism" (p. 28). Here, Snaza (2017) does not suggest abandoning the human, but indeed, accepting the humanness in its animality and relationality with human and nonhuman others. Similarly, Rautio (2013a) argues for an "articulation of interspecies co-existence [which] is the articulation of both our distinctiveness and our interrelations simultaneously" (p. 455). It is noted that the research described so far does lean into the Western minority world experiences, research and theories, of which I am a part and co-contribute quite frequently in a malaise of the current condition of nature juxtaposed to that of the past. Malone (2018) emphasises that "encounters in the new nature movement are often recalled nostalgically with little reference to the diversity of 'childhoods' experienced by children throughout the world" (p. 218). Childhoodnature challenges spacetime by offering a space to accommodate the unknown extremities of childhood multiplicities.

Childhoodnature adopts spacetime conceptualisations that consider how current practices are informed by the multiplicity of past, present, and future activities where linear time is delineated. As such, past ideas of childhood are not retold as childhoodnature but through an entanglement of past, present, and future realities, as is the process of the world's becoming that ideas are (re)turned and (re)turned again: never truly new but never truly the same as spacetime contextualisation matters (Barad, 2014; Malone, 2020). Rousell and Cutter-Mackenzie-Knowles (2020) describe this as "every moment in the life of a child is an uncommon and unrepeatable occasion through which the common world of nature is felt, perceived, and experienced differently" (p. 2), which therefore contributes to "the ongoing reconfigurings of the world" (Barad, 2007, p. 141). Childhoodnature frames possibilities and potentialities, not in preparation or worry about unknown futures, but with the understanding that there is a next moment that matters, and moreover, that this future moment already exists (Barad, 2010; Crinall, 2017).

Drawing specifically upon the childhoodnature literature, the conceptualisations of childhoodnature that have had emphasis in this study are that: Childhoodnature is about challenging: (i) binaries and dualisms, such as mind-body dualisms (Eddy & Moradian, 2018); (ii) the child as more knowledgeable than nonhuman other (Vladimirova & Rautio, 2020); and (iii) human privileging and human-centred orientations (Blenkinsop et al., 2020; Dyment & Green, 2020). Childhoodnature is about childhood *in* nature such as time spent in wild nature (Blenkinsop et al., 2020; Charles & Louv, 2020; Dyment & Green, 2020) and nature play opportunities (Dyment & Green, 2020; Eddy & Moradian, 2018). Childhoodnature is nature *as* teacher (Blenkinsop et al., 2020; Vladimirova & Rautio, 2020). Childhoodnature is child *as* nature (Blenkinsop et al., 2020; Charles & Louv, 2020).

Childhoodnature is a posthuman concept with tenets that inform the theoretical framework of this study (see Figure 3.1). As has been discussed previously, posthuman ways of conceptualising children promote the agency of all matter, that is, the relationality of all material bodies, both human and nonhuman (Malone et al., 2020; Murris, 2020b; Rautio, 2013b; Somerville, 2020). This notion transcends the idea that there are seen boundaries of individualised bodies and moves to (re)consider the continuity of matter (see Barad, 2007). This further enables a reconceptualisation of childhood from being an isolated event in spacetime to an ongoing intra-action with other bodies experiencing the same processes. As Rousell and Cutter-Mackenzie-Knowles (2020) put forward, "everything that a child experiences contributes to the continuity of nature while at the same time irrevocably changing what nature can be"

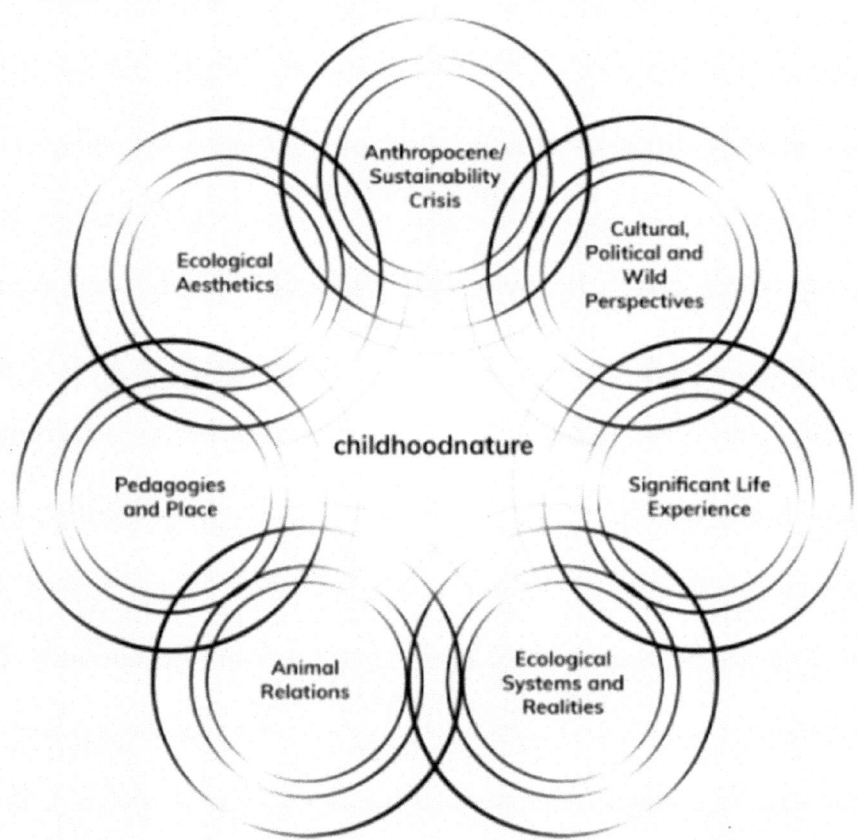

FIGURE 3.1   The childhoodnature model as described through the sections of the *Research Handbook of Childhoodnature* (Cutter-Mackenzie-Knowles et al., 2020). *Illustration by Kelly @ Kelly Designs for the author.*

(p. 9). Somerville (2020) adds that "different bodies of matter mutually change and alter in their ongoing intra-actions" (p. 107). These ideas demarcate childhoodnature as being relational and mutually implicated to align with posthuman thinking where no hierarchical relationship exists between adult/child, parent/child. The agential child is the agential adult as both have equal rights to voice, be heard and be respected on the social and political stage. Utilising childhoodnature bequests a reconsideration of how children and childhood are given these opportunities.

Returning Learning theoretical framework

Our meta/physics, like all good scientific theories, should be alive, responsible, and responsive to the world. How else will our theories matter?

Barad (2011, p. 451)

Materialising from the entanglement of the three concepts defined so far in this Turn – material-discursive practices, affective atmospheres, and childhoodnature – is the Returning Learning theoretical position that frames the work in this book (see Figure 3.2). The Returning Learning theoretical framing provides a rigorous foundation for explaining everyday happenings and encounters where everybody[1] matters. The triage of concepts is theoretically robust each in their own right in disrupting traditional binary-making practices, questioning human exceptionalism, and bringing attention to the mattering of everything particularly the agency of human/nonhuman nature, equally so. Together, their conceptual alignment in these matters provided a fierce theoretical frame to materialise the data entanglements produced in this book.

Although childhoodnature, affective atmospheres, and material-discursive practices are fitting concepts to entangle data through this transqualitative study, the reliance on "naming" that these concepts call for is acknowledged, and troubled (Murris & Haynes, 2018a). The use of language in describing these concepts and the data entanglements that are made is problematised, for example, how one person describes an atmosphere might be different to another. It is extremely subjective and informed by and through the perspective

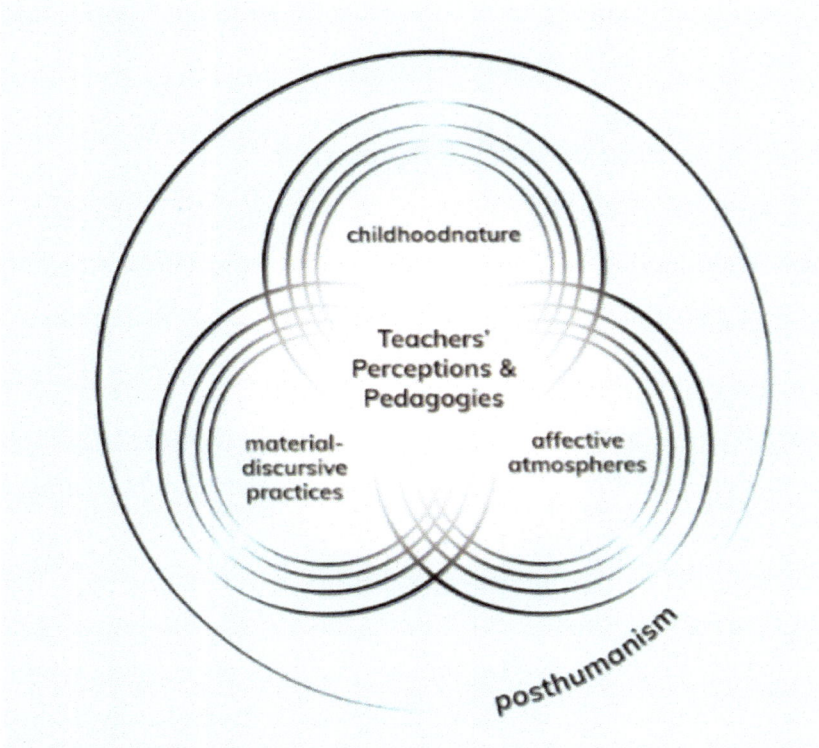

FIGURE 3.2   The Returning Learning theoretical framework espouses an entanglement of childhoodnature, material-discursive practices, and affective atmospheres. *Illustration by Kelly @ Kelly Designs for the author.*

of the viewer. While this may be considered a limitation in some research paradigms, it was embraced in this transqualitative inquiry to demonstrate the richness and depth of data entanglement that is possible when seen through the eyes of the researcher.

Conclusion

Many of the tenets of posthuman that challenged the human-centric and human-dominated model, were attributed to predecessors such as First Nations Peoples, the Romantics, poststructuralists, and feminist theorists. This study adopted cultural posthumanism that follows the lines of thought of Haraway, Braidotti, and Barad.

Posthumanism problematises binary-thinking, argues for a deconstruction of disciplinary silos, and challenges human exceptionalism. Instead, posthumanism proposes a theoretical position that shifts the focus of the attention to include the materiality of human/nonhuman. Through this shift, greater response-ability is enacted for the human in adopting the human body in relationality to all others equally. Through this process of accepting the human body, a "staying-with" the tensions of posthumanism was enacted in responding to the global crises of this spacetime.

This Turn has explored how *matter extends beyond the physical and has agency that communicates* through the concept of material-discursive practices. Material-discursive practices espouse the agency of matter – the "voice" of all human/nonhuman materiality. These practices extend beyond form and as such, demonstrate a deconstruction of binary-making efforts that push the human and nonhuman into distinctively separate realms. Material-discursive practices shift the human-centred focus to a redistribution across the human/nonhuman domain. As such, materiality then is considered the seen/unseen, material-discursive forces that constitute its making. The significance of material-discursive practice is its potential to explore the "behind-the-seens" intra-actions and contributions that are always already being made in and out of classrooms by early school years teachers' perceptions and pedagogies.

I further investigated the theoretical offering of how *physical spaces hold a quality that can be felt* through affective atmospheres. Affective atmospheres describe the mood, vibe, and background intensities of spaces that make themselves known in sometimes discrete, or even conspicuous ways. Affect is defined separately from emotion and as such, is not subjective but an intensity. This affectivity is applied in affective atmospheres, which explain the material dimensions of the human/nonhuman in intra-action. Through its very nature, affective atmospheres challenge human exceptionalism by bringing nonhuman agents equally to the forefront of human perception. Moreover, binaries are challenged through understanding the relationality of space, where dualisms do not exist. These background affectivities create atmospheres that are necessary to understand in educational settings such as inside the classroom or outside in nonhuman nature to enact change through embodied, visceral responses.

Finally, *binary-making practices are exposed and troubled* as a theoretical offering through the concept of childhoodnature. Childhoodnature proposes a conceptual understanding of childhood and nature being intra-actively constituted such that there is no boundary between them. Childhoodnature argues against child/nature binaries and acts, as the posthuman child, to deconstruct practices of misopedy that enforce the empowerment of adult over child. Moreover, childhoodnature attends to practices that define the child as separate from nature; the child is inherently a part of nature when viewed through this lens.

These three concepts are synthesised to form the Returning Learning theoretical framework which was used to think with and through the diffractive data entanglements that are presented in the Eighth and Ninth Turns.

Note

1 A body here refers to any entity or material object. This could be an atom, a rock, a tree, a person, or a building.

References

Alaimo, S. (2010). *Bodily natures: Science, environment, and the material self.* Indiana University Press.

Änggård, E. (2016). How matter comes to matter in children's nature play: Posthumanist approaches and children's geographies. *Children's Geographies, 14*(1), 77–90. https://doi.org/10.1080/14733285.2015.1004523

Ash, J., & Anderson, B. (2015). Atmospheric methods. In P. Vannini (Ed.), *Non-representational methodologies* (pp. 44–61). Routledge.

Bai, H., Elza, D., Kovacs, P., & Romanycia, S. (2010). Re-searching and re-storying the complex and complicated relationship of biophilia and bibliophilia. *Environmental Education Research, 16*, 3–4. https://doi.org/10.1080/13504621003613053

Barad, K. (2007). *Meeting the universe halfway: Quantum physics and the entanglement of matter and meaning.* Duke University Press.

Barad, K. (2010). Quantum entanglements and hauntological relations of inheritance: Dis/continuities, spacetime enfoldings, and justice-to-come. *Derrida Today, 3*(2), 240–268.

Barad, K. (2011). Erasers and erasures: Pinch's unfortunate "uncertainty principle". *Social Studies of Science, 41*(3), 443–454.

Barad, K. (2012). On touching – The inhuman that therefore I am. *Differences, 23*(3), 206–223.

Barad, K. (2014). Diffracting diffraction: Cutting together-apart. *Parallax, 20*(3), 168–187. https://doi.org/10.1080/13534645.2014.927623

Bissell, D. (2010). Passenger mobilities: Affective atmospheres and the sociality of public transport. *Environment and Planning D: Society and Space, 28*(2), 270–289.

Blackman, L., & Venn, C. (2010). Affect. *Body & Society, 16*(1), 7–28.

Blaikie, F., Daigle, C., & Vasseur, L. (2020). *New pathways for teaching and learning: The posthumanist approach.* Canadian Commission for UNESCO.

Blaise, M. (2013). Activating micropolitical practices in the early years (re)assembling bodies and participant observations. In R. Coleman & J. Ringrose (Eds.), *Deleuze and research methodologies* (pp. 184–200). Edinburgh University Press. http://www.jstor.org/stable/10.3366/j.ctt1g0b6xx.14

Blenkinsop, S., Jickling, B., Morse, M., & Jensen, A. (2020). Wild pedagogies: Six touchstones for childhoodnature theory and practice. In A. Cutter-Mackenzie, K. Malone, & E. Barratt Hacking (Eds.), *Research handbook on childhoodnature: Assemblages of childhood and nature research* (pp. 1–18). Springer International Publishing.

Braidotti, R. (2013). *The posthuman*. Polity Press.

Braidotti, R., & Dolphijn, R. (2017). *Philosophy after nature*. Rowman & Littlefield International.

Carson, R. (1965). *Sense of wonder*. Harper & Row.

Castree, N., & Nash, C. (2006). Posthuman geographies. *Social & Cultural Geography, 7*(4), 501–504. https://doi.org/10.1080/14649360600825620

Castree, N., Nash, C., Badmington, N., Braun, B., Murdoch, J., & Whatmore, S. (2004). Mapping posthumanism: An exchange. *Environment and Planning A, 36*(8), 1341–1363.

Charles, C., & Louv, R. (2020). Wild hope: The transformative power of children engaging with nature. *Research handbook on childhoodnature: Assemblages of childhood and nature research* (pp. 1–21). Springer International Publishing.

Crinall, S. (2017). Bodyplacetime: Painting and blogging "dirty, messy" human-natured becomings. In K. Malone, S. Truong, & T. Gray (Eds.), *Reimagining sustainability in precarious times* (pp. 95–114). Springer. https://doi.org/10.1007/978-981-10-2550-1_7

Cutter-Mackenzie-Knowles, A., Brown, S. L., Osborn, M., Blom, S. M., Brown, A., & Wijesinghe, T. (2020a). Staying-with the traces: Mapping-making posthuman and indigenist philosophy in environmental education research. *Australian Journal of Environmental Education, 36*(2), 105–128.

Cutter-Mackenzie-Knowles, A., Malone, K., & Barratt Hacking, E. (2020b). *Research handbook on childhoodnature: Assemblages of childhood and nature research* (A. Cutter-Mackenzie, K. Malone, & E. Barratt Hacking, Eds.). Springer International Publishing.

Dernikos, B. P., Lesko, N., McCall, S. D., & Niccolini, A. D. (2020a). Feeling education. In *Mapping the affective turn in education* (pp. 3–27). Routledge.

Dernikos, B. P., Lesko, N., McCall, S. D., & Niccolini, A. D. (2020b). *Mapping the affective turn in education: Theory, research, and pedagogies*. Routledge.

Dewey, J. (1916). *Democracy and education: An introduction to the philosophy of education*. The Macmillan Company.

Dyment, J., & Green, M. (2020). Everyday, local, nearby, healthy childhoodnature settings as sites for promoting children's health and well-being. In A. Cutter-Mackenzie-Knowles, K. Malone, & E. Barratt Hacking (Eds.), *Research handbook on childhoodnature: Assemblages of childhood and nature research* (pp. 1–26). Springer International Publishing.

Eddy, M. H., & Moradian, A. L. (2018). Childhoodnature in motion: The ground for learning. In A. Cutter-Mackenzie-Knowles, K. Malone, & E. Barratt Hacking (Eds.), *Research handbook on childhoodnature: Assemblages of childhood and nature research* (pp. 1–24). Springer International Publishing.

Finn, M. (2016). Atmospheres of progress in a data-based school. *Cultural Geographies*, *23*(1), 29–49.

Haraway, D. (1992). The promises of monsters: A regenerative politics for inappropriate/d others. *Cultural Studies*, 295–337. https://www.professores.uff.br/ricardobasbaum/wp-content/uploads/sites/164/2023/04/Haraway_The_Promise_of_Monsters_1992.pdf

Haraway, D. J. (2016). *Staying with the trouble: Making kin in the Chthulucene*. Duke University Press.

Jackson, A. Y., & Mazzei, L. (2012). *Thinking with theory in qualitative research: Viewing data across multiple perspectives*. Routledge.

Jansen, Y., Leeuwenkamp, J., & Urricelqui, L. (2021). Post-everything. In *Posthumanism and the "posterizing impulse"*. Manchester University Press. https://doi.org/10.7765/9781526148179.00020

Kalvaitis, D., & Monhardt, R. (2015). Children voice biophilia: The phenomenology of being in love with nature. *The Journal of Sustainability Education*, *9*. https://www.susted.com/wordpress/content/children-voice-biophilia-the-phenomenology-of-being-in-love-with-nature_2015_03/

Latour, B. (2004). *The politics of nature*. Harvard University Press.

Lenz Taguchi, H. (2010). *Going beyond the theory/practice divide in early childhood education: Introducing an intra-active pedagogy*. Routledge.

Lorimer, J. (2009). Posthumanism/Posthumanistic geographies. In R. Kitchin & N. Thrift (Eds.), *International encyclopedia of human geography* (pp. 344–354). Elsevier. https://doi.org/10.1016/B978-008044910-4.00723-9

Loughland, T., Reid, A., & Petocz, P. (2002). Young People's conceptions of environment: A phenomenographic analysis. *Environmental Education Research, 8*(2). https://doi.org/10.1080/13504620220128248

Louv, R. (2006). *Last child in the woods: Saving our children from nature-deficit disorder*. Atlantic Books.

MacKenzie, I. (2001). Unravelling the knots: Post-structuralism and other "Post-isms". *Journal of Political Ideologies, 6*(3), 331–345.

Malone, K. (2018). *Children in the Anthropocene: Rethinking sustainability and child friendliness in cities*. Palgrave Macmillan, Springer Nature.

Malone, K. (2020). Re-turning childhoodnature: A diffractive account of the past tracings of childhoodnature as a series of theoretical turns. *Research handbook on childhoodnature* (pp. 1–31). Springer International Publishing.

Malone, K., Duhn, I., & Tesar, M. (2020). Greedy bags of childhoodnature theories. In A. Cutter-Mackenzie-Knowles, K. Malone, & E. Barratt Hacking (Eds.), *Research handbook on childhoodnature*. Springer International Publishing.

Massumi, B. (1995). The autonomy of affect. *Cultural Critique, 31*, 83–109.

Mayes, E., Wolfe, M. J., & Higham, L. (2020). Re/imagining school climate: Towards processual accounts of affective ecologies of schooling. *Emotion, Space and Society, 36*, 100703.

Menning, S. F., Murris, K., & Wargo, J. M. (2020). Reanimating video and sound in research practices. *Navigating the postqualitative, new materialist and critical posthumanist terrain across disciplines* (pp. 150–168). Routledge.

Merrell, F. (2003). *Sensing corporeally: Toward a posthuman understanding*. University of Toronto Press.

Miah, A. (2008). A critical history of posthumanism. *Medical enhancement and posthumanity* (pp. 71–94). Springer.

Murris, K. (2016). *The posthuman child: Educational transformation through philosophy with picturebooks*. Routledge. https://doi.org/10.4324/9781315718002

Murris, K. (2018). Posthumanism, de/colonising education and child (hoods) in South Africa. *Literacies, literature and learning* (pp. 25–49). Routledge.

Murris, K. (2020a). The "missing peoples" of critical posthumanism and new materialism. In *Navigating the postqualitative, new materialist and critical posthumanist terrain across disciplines* (pp. 62–84). Routledge.

Murris, K. (2020b). Posthuman child and the diffractive teacher: Decolonizing the Nature/Culture binary. In A. Cutter-Mackenzie, K. Malone, & E. Barratt Hacking (Eds.), *Research handbook on childhoodnature: Assemblages of childhood and nature research* (pp. 1–25). Springer International Publishing. https://doi.org/10.1007/978-3-319-51949-4_7-2

Murris, K. (2022). *Karen barad as educator: Agential realism and education*. Springer Singapore Pte. Limited.

Murris, K., & Borcherds, C. (2019). Body as transformer: "Teaching without teaching" in a teacher education course. *Posthumanism and higher education* (pp. 255–277). Springer.

Murris, K., Crowther, J., & Stanley, S. (2018). Digging and diving for treasure: Erasures, silences and secrets. *Literacies, literature and learning* (pp. 149–172). Routledge.

Murris, K., & Haynes, J. (2018a). *Literacies, literature and learning: Reading Classrooms differently*. Routledge.

Murris, K., & Haynes, J. (2018b). Philosophical playthinking in a South African literacy "classroom". *Literacies, literature and learning* (pp. 3–24). Routledge.

O'Neil, J. K. (2018). Transformative sustainability learning within a material-discursive ontology. *Journal of Transformative Education*, *16*(4), 365–387.

Petitfils, B. (2014). Researching the posthuman: The "subject" as curricular lens. *Posthumanism and educational research* (pp. 30–42). Routledge.

Prout, A. (2011). Taking a step away from modernity: Reconsidering the new sociology of childhood. *Global Studies of Childhood*, *1*(1), 4–14.

Rautio, P. (2013a). Being nature: Interspecies articulation as a species-specific practice of relating to environment. *Environmental Education Research*, *19*(4), 445–457. https://doi.org/10.1080/13504622.2012.700698

Rautio, P. (2013b). Children who carry stones in their pockets: On autotelic material practices in everyday life. *Children's Geographies*,
11(4), 394–408.

Rousell, D., & Cutter-Mackenzie-Knowles, A. (2020). Uncommon
worlds: Toward an ecological aesthetics of childhood in the Anthropocene. In A. Cutter-Mackenzie-Knowles, K. Malone, & E. Barratt
Hacking (Eds.), *Research handbook on childhoodnature: Assemblages of childhood and nature research* (pp. 1–23). Springer International Publishing. https://doi.org/10.1007/978-3-319-51949-4_88-2

Snaza, N. (2013). Bewildering education. *Journal of Curriculum and Pedagogy*, *10*(1), 38–54. https://doi.org/10.1080/15505170.2013.783889

Snaza, N. (2014). Toward a genealogy of educational humanism. *Posthumanism and educational research* (pp. 17–29). Routledge.

Snaza, N. (2017). Is john Dewey's thought" humanist"? *JCT (Online)*,
32(2), 15–34.

Snaza, N. (2020). Love and bewilderment: On education as affective
encounter. *Mapping the affective turn in education* (pp. 108–121).
Routledge.

Snaza, N., Appelbaum, P., Bayne, S., Carlson, D., Morris, M., Rotas,
N., Sandlin, J., Wallin, J., & Weaver, J. A. (2014). Toward a posthuman education. *Journal of Curriculum Theorizing*, *30*(2), 39.

Snaza, N., Sonu, D., Truman, S. E., & Zaliwska, Z. (2016). *Pedagogical matters: New materialisms and curriculum studies*. Peter Lang
Publishing, Inc.

Snaza, N., & Weaver, J. (2014). *Posthumanism and educational research*. Routledge.

Somerville, M. (2020). Posthuman theory and practice in early years
learning. In A. Cutter-Mackenzie, K. Malone, & E. Barratt Hacking
(Eds.), *Research handbook on childhoodnature: Assemblages of
childhood and nature research* (pp. 1–25). Springer International
Publishing. https://doi.org/10.1007/978-3-319-51949-4_6-1

Taylor, C. A. (2013). Objects, bodies and space: Gender and embodied practices of mattering in the classroom. *Gender and Education*,
25(6), 688–703. https://doi.org/10.1080/09540253.2013.834864

Taylor, C. A. (2016). Close encounters of a critical kind: A diffractive
musing in/between new material feminism and object-oriented ontology. *Cultural Studies? Critical Methodologies*, *16*(2), 201–212.

Taylor, C. A. (2020). Knowledge matters: Five propositions concerning the reconceptualisation of knowledge in feminist new materialist, posthumanist and postqualitative approaches. *Navigating the postqualitative, new materialist and critical posthumanist terrain across disciplines* (pp. 22–42). Routledge.

Taylor, C. A., Quinn, J., & Franklin-Phipps, A. (2020). Rethinking research "use": Reframing impact, engagement and activism with feminist new materialist, posthumanist and postqualitative research. *Navigating the postqualitative, new materialist and critical posthumanist terrain across disciplines* (pp. 169–189). Routledge.

Thrift, N. (2008). *Non-representational theory: Space, politics, affect.* Routledge.

Tuana, N. (2008). Viscous porosity: Witnessing Katrina. In S. Alaimo & S. Hekman (Eds.), *Material feminisms* (p. 434). Indiana University Press.

Verlie, B. (2018). *Affective entanglements: learning to live-with climate change* [PhD Diss., Monash University].

Verlie, B., & Blom, S. M. (2021). Education in a changing climate: Reconceptualising school and classroom climate through the fiery atmos-fears of Australia's Black summer. *Children's Geographies*, 1–15. https://doi.org/10.1080/14733285.2021.1948504

Vladimirova, A., & Rautio, P. (2020). Unplanning research with a curious practice methodology: Emergence of childrenforest in the context of Finland. *Research handbook on childhoodnature: Assemblages of childhood and nature research* (pp. 1–26). Springer International Publishing.

Wolfe, M. J., & Rasmussen, M. L. (2020). The affective matter of (Australian) school uniforms: The school-dress that is and does. *Mapping the affective turn in education* (pp. 179–193). Routledge.

Zeilinger, A. (2010). Quantum physics: Ontology or epistemology? *The Trinity and an entangled world: Relationality in physical science and theology* (pp. 32–40). Wiley.

Zembylas, M. (2020). The affective atmospheres of democratic education: Pedagogical and political implications for challenging right-wing populism. *Discourse: Studies in the Cultural Politics of Education*, *43*(1), 1–15.

Zembylas, M. (2021). The affective turn in educational theory. *Oxford Research Encyclopedia of education*. Oxford Research Encyclopedia.

4

THE FOURTH TURN

Methodologies that matter for future making in educational research

Introduction

I want to begin by re-turning – not by returning as in reflecting on or going back to a past that was, but re-turning as in turning it over and over again – iteratively intra-acting, re-diffracting, diffracting anew, in the making of new temporalities (spacetimematterings), new diffraction patterns. We might imagine re-turning as a multiplicity of processes, such as the kinds earthworms revel in while helping to make compost or otherwise being busy at work and at play: turning the soil over and over – ingesting and excreting it, tunnelling through it, burrowing, all means of aerating the soil, allowing oxygen in, opening it up and breathing new life into it. It might seem a bit odd to enlist an organic metaphor to talk about diffraction, an optical phenomenon that might seem lifeless. But diffraction is not only a lively affair, but one that troubles dichotomies, including some of the most sedimented and stabilized/stabilizing binaries, such as organic/inorganic and animate/inanimate. Indeed, the quantum understanding of diffraction

DOI: 10.4324/9781032703473-4

troubles the very notion of dicho-tomy – cutting into two – as a singular act of absolute differentiation, fracturing this from that, now from then.

Re-turning as a mode of intra-acting with diffraction – diffracting diffraction – is particularly apt since the temporality of re-turning is integral to the phenomenon of diffraction.

Barad (2014, p. 168)

Barad (2014) aptly opens this Turn to provide the conceptual and theoretical foundation for the diffractive approach under-pinning the "data entanglements" (the term used to describe the phenomenon of indeterminate data collection and data analysis practices) proposed in this book. More on diffractive data entan-glements to come.

As described in previous Turns, the purpose of this book is to explore how early school years teachers perceive childhoodna-ture and how it informs their pedagogy through a posthuman lens using the Returning Learning theoretical framework (see the Third Turn). In line with this theoretical positioning, the research was undertaken using a diffractive ethnographic approach and a transqualitative methodology. This methodology in combination with this methodological approach has not previously been ap-plied to educational research providing an exciting, novel way to undertake empirical research and consider alternative ap-proaches to the normalised "business-as-usual" practices for a planet and a people that desperately need it.

Drawing on the Baradian ethico-onto-epistemology proposed in the Third Turn, transqualitative inquiry is proposed in this book to move beyond conventional qualitative research and provide an alternative to post qualitative inquiry (PQI). Transqualitative inquiry is explored and defined later in this Turn. The approach adopted to enact a transqualitative methodology is diffractive ethnography which is nested within the methodological positioning of this study and aligns to the Returning Learning theoretical framework. Through the application of transqualitative inquiry into practice, data is reconceptualised, and "done" differently as diffractive data entanglements.

A note on diffractive data entanglements...

Because entanglements are vital to understanding relationalities in the specificities, what is needed is an approach that does not tear them apart....

Barad & Gandorfer (2021, p. 51)

What we (researchers, scholars, students, teachers, etc.) do with "data" once we have "access" to it happens often unexpectedly, in unpredictable and entangled ways.

Koro-Ljungberg (2015, p. 48)

The concept of entanglement explains the way matter intra-acts. For the purpose of this book, the concept of entanglements is used to describe the way data makes itself known. That is, data collection and data analysis are re-framed as data entanglements to signal the constant, inherent intra-actions that are always already happening between data collection and analysis practices such that their distinction is indeterminable.

Entanglements are the meeting of material-discursive practices through their diffractive patterning that creates disturbances. These enactments cannot be separated or drawn apart but are part of material becoming. The material becoming is the way that a phenomena, in this case the data, makes itself known. Data entanglements are not merely "the joining of separate entities" (Barad, 2007, p. ix), such as data collection and analysis, but describe a dynamism – an inherent understanding that these data-making practices are in constant flux. Entanglement does not denote a blending or unification: "entanglements are not unities. They do not erase differences; on the contrary, entanglings entail differentiatings, differentiatings entail entanglings. One move – *cutting together-apart*" (Barad, 2014, p. 176; emphasis in original). An entanglement describes the relationality of matter as always already part of a collective process of becoming. The data entanglements in this study are inherently relational and make themselves known through the research processes that were enacted.

In addition, entanglements deconstruct notions of the self, or "I" by positioning the human as a dynamic phenomenon with others (Barad, 2007; Koro-Ljungberg, 2015; St. Pierre, 2011). As stated by Snaza et al. (2016) "there is no longer a knowing (human) subject who acts and a passive (nonhuman) object that is acted upon: everything is 'entangled'" (p. xvii). St. Pierre (2013) argues that "if being is always already entangled, then something called data cannot be separate from me, 'out there' for 'me' to 'collect'" (p. 226) and continues to justify the discontinued need for data as a defined "thing". However, I contend that data is still a needed concept albeit one that requires reconceptualisation. I concur with St. Pierre (2013) that the notion of data being "out there" for "me" to "collect" is problematic, and as such, I explore ways that data is always already entangled in the data-making process. That is, I argue that data are not "passive" or "subservient to the work of analysis and interpretation", nor are data "inert, lifeless and disorganized" (Koro-Ljungberg et al., 2018, pp. 462–463). Data are material, whereas data as matter is "lively, agentic, and infused with affect" (Koro-Ljungberg et al., 2018, p. 469). As Koro-Ljungberg (2015, p. 46) describes,

> Thinking about data's relationality, movement, entanglements, or multidirectional epistemological flows – that is, knowledge from data shaping researchers and research, knowledge from research shaping data, and/or knowledge within the data-researcher relationship shaping the data-researcher relationship, among others – might help us to change the direction of knowledge production in critical social science research and practice.

While the word entanglement has often been used interchangeably with assemblage in the literature (see for example Ellingson & Sotirin, 2020; Gullion, 2018), in this book it is necessary to clarify that an entanglement is a term from quantum theory to describe the science that explains how matter cannot strictly be defined or separated with boundaries (Barad, 2007), although the distinctions may seem *real* to the human eye. The two concepts of entanglement and assemblage could be argued and conceptualised to find the intersections, as was undertaken by MacLure (2013) who describes assemblage and entanglement to explain "non-hierarchical organisation". However, the work undertaken in this book is grounded and positioned in the science of quantum physics by considering the human/nonhuman intra-actions of the classroom milieu such as through the concept of material-discursive practices. In addition, quantum theory is where the concept of entanglements was born and as such, is applied fittingly here.

Diffractive data entanglements additionally offer a space to dwell in the tension that is experienced when ethnographic data methods are applied to a posthuman framing. To this end, Davies (2021) elucidates how "the concept of entanglement disrupts the taken-for-granted ascendancy of the human species, and it abandons the humanist version of what we are – moving from identities to emergent, singular multiplicities" (p. 140). The tension lies in thinking with concepts that problematise human exceptionalism and the "taken-for-granted ascendancy of the human species" as Davies (2021, p. 140) states above, while engaging in research practices that are always already human centred as it is the I that is doing the research. Taylor (2020) concurs, stating that "this reconceptualisation of "I" as a relational emergence within entangled "practices of knowing in being" (Jackson & Mazzei, 2011, p. 116) brings ethical issues concerning values (positionality) and response-ability (accountability)" (p. 36). Diffractive data entanglements are informed through an ethico-onto-epistemology position proposed by Barad (2007; see the Third Turn), where responsibility is a response-ability to "attending to, tracing, and taking account of entanglements, about being in touch with world's practices of materialising/making-sense, including its material-discursive 'concepting'" (Barad & Gandorfer, 2021, p. 31). Therefore, the relationality of entanglements is highlighted where "securing objectivity in research is not about disentangling and disengaging the subject from the object (as in much research), but "taking responsibility for one's entanglements" (Barad, 2007, p. 453, footnote 1)" (Murris, 2020a, p. 8).

Ellingson and Sotirin (2020) promote the use of the term "data engagements" as a "generative alternative" (p. 817) to refute the abolishment of the concept of data altogether as is oft proposed in PQI (Denzin, 2013; St. Pierre, 2013). Data engagements involve the processes of making, assembling and becoming data (Ellingson & Sotirin, 2020). While Ellingson and Sotirin (2020) invite all colleagues "across the methodological continuum" to join in the use of data engagements (p. 824), I contend that the concept of data engagements does not align with the ethico-onto-epistemological theoretical framework of this study as it espouses a consideration of data as assemblage – a different contention and position than a diffractive data entanglement.

However, I do draw on Ellingson and Sotirin's (2020) concept of "data making" to refer to the diffractive data entanglement process. Moreover, I highlight that *data making* is distinctly different from *research-creation* (Truman, 2023): the former is about bringing data into being (Ellingson & Sotirin, 2020) and the latter is concerned with bringing something into existence (Loveless, 2019; Truman, 2023). The distinction is that *data making* implies that "data may become data simply by labelling and curating them as such" (Ellingson & Sotirin, 2020, p. 820) and can involve a range of creative and innovative ways of doing so. Whereas, *research-creation* generally involves artful practices that literally bring data into existence; data that was not there prior to the research processes taking place (Loveless, 2019; Sweet et al., 2020). I make the data in the process that is akin to the interventions proposed by Verlie (2019) or through a quantum theory lens, which can be likened to an agential cut (Barad, 2014) where I, as the researcher, purposefully choose the way the data will be read to generate "new" data diffractions from the original data set.

Data entanglements as an agential cut through diffraction gratings

> At the same time as separating out or excluding "something" – an object, practice, person, an event – from the ongoing flow of spacetimemattering, cuts entangle us ontologically with/in and as the phenomena produced by the cut we make. This is what Barad (2007) means by cutting together-apart. You are entangled as knowledge-ing materialises; you are co-implicated and therefore accountable because knowledge-making is intra-relational, emergent, situated and contingent … knowledge is an ethico-onto-epistemological practice, which (a) entangles us as researchers in matters of accountability and response-ability, and (b) which shapes the knowledge pro-duced through research and pedagogy as a material practice which entails cuts that enact judgements and produce differ-ential matterings.
>
> Taylor (2020, p. 37)

The diffractive act of making data entanglements involves pushing the data through a series of diffraction gratings which act as agential cuts into the data. As described above by Taylor (2020), an agential cut is a boundary-making practice where the influence and decision-making practices of the researcher are made (Barad, 2007).

The process of using diffraction gratings is demonstrated in Figure 4.1. Through the first diffraction grating, data was entan-gled with the concept of childhoodnature while simultaneously being pushed through the concepts of affective atmospheres and material-discursive practices (for an elaboration of these con-cepts, see the Third Turn) which also act as "diffraction gratings"

(Barad, 2007, p. 78) to generate diffractive patterning (see Figure 4.1). This model depicted in Figure 4.1 is based on the two-slit experiment initially proposed by Niels Bohr and later developed by Marlan Scully and his associates (see Barad, 2007, p. 307). Figure 4.1 explains the iterative making of diffractive data entanglements: the process where the data makes itself known to the researcher, the researcher engages with that data and documents the data through the theoretical framework.

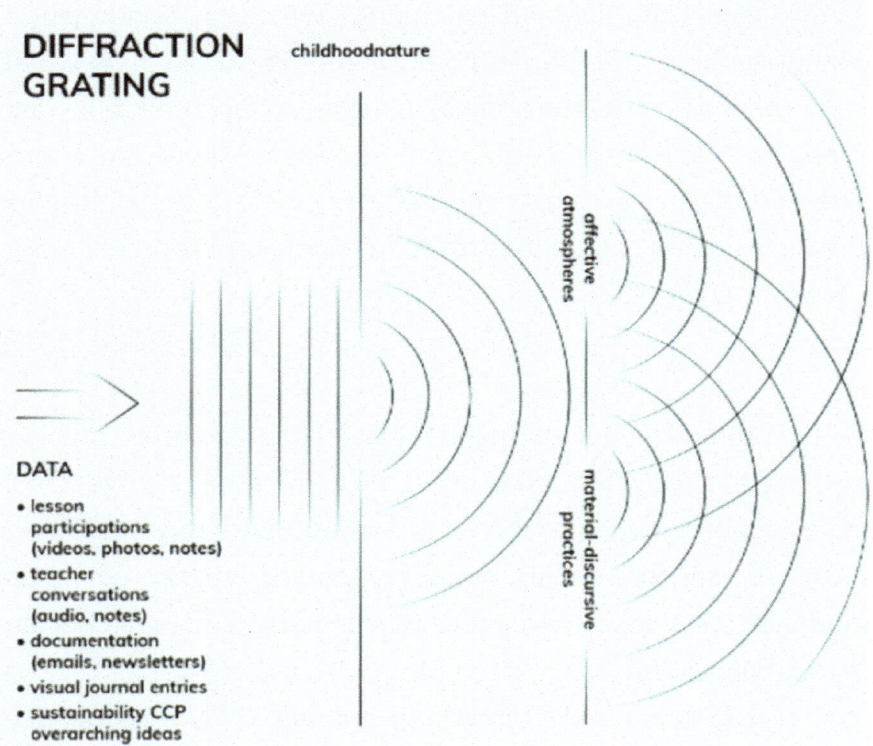

FIGURE 4.1 The diffraction grating created by the concepts of childhoodnature, affective atmospheres and material-discursive forces. Image inspired by the work of Barad (2007; see Figure 4, p. 78 and Figure 5, p. 79). *Illustration by Kelly @ Kelly Designs for the author.*

From this discussion of challenging conventional practices of data collection and data analysis, it is evident that it is no longer appropriate to describe research processes so separately and determinately from a posthuman, ethico-onto-epistemological position (Ellingson & Sotirin, 2020; St. Pierre, 2014a). In light of this, the research in this book arrived at the concept of diffractive data entanglements to describe the dynamic, fluid, iterative and entangled process of data collection and analysis.

Methodology making

As discussed in the introduction of this Turn, this book describes a transqualitative methodology to position the diffractive data entanglements within the posthuman theories underpinning this research. For this to happen, it was necessary to understand why a "new" and different methodology was required by exploring the need to push beyond both qualitative and post-qualitative methodologies and that exploration is presented here.

Qualitative research

At its core, qualitative research is a "situated activity that locates the observer in the world" (Denzin & Lincoln, 2008, p. 4). Put simply, it seeks to understand social settings by placing the observer *in* the research. Qualitative research is an overarching research methodology adopted largely by the humanities and social sciences (Denzin & Lincoln, 2008; Nelson et al., 1992; Silverman, 2016) and seeks to understand the "nature of phenomena" rather than its magnitude and distribution, which characterises quantitative research (Musante & DeWalt, 2010, p. 2). It challenges conventional quantitative methodologies by focussing on the subjective and interpretative nature of the associated methods. These are "material practices that make the world visible" (Denzin & Lincoln, 2008, p. 4).

As so far described, qualitative research is human centric and focuses on human behaviours, interactions and relationships. However, this directive very clearly does not align with posthuman thinking and thus creates a tension in adopting qualitative approaches to posthuman research. In this book, this tension is disrupted by offering a way of doing research that bridges these juxtapositions. It does so by adopting qualitative methods in practice and theorising the data with posthuman concepts. Rather than implementing a wholly conventional qualitative paradigm – where the misalignment in adopting posthumanist thinking through a humanist framework created too much tension – a transqualitative framework emerged. The emergence of the transqualitative was not to undermine the great work done by post qualitative researchers thus far but to acknowledge that the post qualitative did not accommodate some qualitative research traditions. As such, it is the way that qualitative methods are conceptualised, theorised and practised that determines their position and ability to be applied to different contexts, such as in transqualitative research. While transqualitative research (discussed in full later in the Turn) proposes a movement *beyond* conventional qualitative research, the foundations of qualitative research are still visible in trans*qualitative* inquiry by placing the researcher *in* the research and exploring the *nature of phenomena*.

Post qualitative inquiry

PQI as an approach to undertaking research was first proposed by St. Pierre (2011) as a way to enable the thinking together of poststructuralist theories through a research approach that was not set in the "conventional humanist qualitative" paradigm's ideas of "human being, language, discourse, power, agency, resistance, freedom, and so on" (St. Pierre, 2018, p. 603). I will upfront by clarifying that I will use "post qualitative" as two words, unhyphenated, in respect of the post qualitative work of St. Pierre who founded the field and has defined it as such since 2011, and ongoing at the time of writing this publication in 2024.

PQI, as proposed by St. Pierre (2011) "begins with" poststructuralist theory (St. Pierre, 2024, p. 4); that is, PQI was born from the need for a research methodology that enabled poststructuralist thinking to be applied in practice (St. Pierre, 2023). St. Pierre (2018) provides an approach to *doing* PQI that involves engaging deeply with poststructuralist theorists such that the "post qualitative researcher must *live* the theories (will not be able to *not* live them)" (p. 604, emphasis in original).[1] St. Pierre (2023) brings attention to ontology, noting that "epistemology remains dominant; after all, universities are big knowledge-production machines" (p. 22).

PQI is based on a one-world ontology (St. Pierre, 2019) proposed by poststructuralist theory that shifts focus from the what is of *presence* to the what might be of *immanence*. It espouses a flat surface on which there is no difference between the virtual and the actual; they become one and the same. St. Pierre (2019) compares an ontology of immanence to the two-world ontology proposed by social scientists who explore "conditions of possibility" and thematic analytic research practices that assume "internal relations" (p. 6) stating that "concepts in the ontology of immanence and transcendental empiricism which inform post qualitative inquiry" (p. 7) to identify the philosophical, theoretical and ontological position of PQI. These boundaries are elaborated by stating that PQI rejects pilot studies because "immanence cannot be piloted" (p. 8).

PQI has some guides for engagement that involves being informed by and of living poststructural theory that creates "different worlds for living" (St. Pierre, 2018, p. 604). St. Pierre (2024) explains in depth how philosophers such as "Baudrillard, Bergson, Deleuze, Deleuze with Guattari, Derrida, Foucault, Lyotard, and others" (St. Pierre, 2024, p. 4) proposed "new ways to think and live" (St. Pierre, 2024, p. 5). To understand and embody these poststructuralists involves reading – "slow reading ... [and] difficult reading" (St. Pierre, 2024, p. 6) for those interested in PQI.

Thinking with poststructuralist philosophers and theorists such as Derrida, Foucault, Deleuze and Guattari, St. Pierre (2011) proposed a new way of doing research – which St. Pierre (2023) terms "inquiry" – that more aptly aligned poststructural theory to practice, such that poststructuralist theory, which rejects humanism, was not trying to fit into a methodology that is embedded in it. Indeed, through poststructural theory, the notion of methodology was not considered part of PQI at all.

St. Pierre first became aware of the incongruencies of poststructuralist theories with qualitative methodologies during their own PhD study, where they attempted qualitative research methods alongside poststructuralist theories. Moreover, St. Pierre (2018) discovered that qualitative methodology had surprisingly become,

so formalized, systematized and positivized ... given that it was invented during the interpretive turn that resisted positivism and shifted from measurement, quantification, and prediction to understanding people's lived experiences (interpretive research) and, when those experiences seemed unjust, to transformation and liberation (critical research).

p. 603

As more recently noted by St. Pierre (2023), qualitative practices and processes (such as a literature review and data practices) do not align with the onto-epistemology of poststructuralism and to adopt them in PQI would be "nonsensical" (p. 7). This position is held by other post qualitative researchers (Koro, 2021; Mazzei, 2021; Taylor, 2020; Tesar, 2021) who challenge known research processes. As stated by Murris (2020a), "post qualitative research rejects the use of fixed methods and proposes philosophy or concept as method" (p. 3), as Taguchi and St. Pierre (2017) have previously explored "using concept as method in education and social science inquiry" (p. 643). St. Pierre (2023) further claimed qualitative research as "simple.... So, dreadfully boring" (p. 23).

PQI has indeed revolutionised how we, as researchers or "inquirers" (St. Pierre, 2023), consider our practice. It offers researchers alternative approaches to doing research outside of qualitative and quantitative methodologies. The promise of doing research where the "post-qualitative inquirer does not know what to do first and then next and next" where "there is no recipe, no process" (St. Pierre, 2018, p. 604) is appealing to researchers who are adopting creative research processes and who do not want to follow traditional research conventions (for example, MacLure, 2024a; MacLure, 2024b; Murris, 2020b; Murris & Kohan, 2021). This is worthy research as it extends and expands on what is known in ways that are unknown, uncertain, and risky as demonstrated, for example, by the works described in this Turn along with the other books published in this series. I love that researchers/inquirers have found their space to "play" and conduct their inquiry through the post qualitative – enjoying the limitless possibilities and enabling freedom that this offers (Murris, 2020a). Such as Murris (2020a) in the opening chapter of the introductory text to this series where researcher authors were encouraged to use terminologies of familiarity to facilitate ease of expression rather than "reducing/translating/homogenising" it. I recently searched Google Scholar for "post-qualitative" since 2023, and returned 1780 results. It confirmed the burgeoning of the field, and how desperately researchers have been calling for alternative approaches to engaging in research as inquiry guided by theory.

Although PQI resists definition (Aagaard, 2022; Lester, 2021), Lester (2021) views PQI as an *invitation* to experiment, to sit with theory and to consider the "next and next" (as proposed by St. Pierre, 2018). Aagaard (2022) defines it through three tenets: "relational ontologies, new materialism and troubling methods" (p. 313). I agree that post qualitative troubles methods and methodologies, indeed, this is how the field was founded. However, I would argue that PQI resides in poststructuralist theories over new materialism as one of the three premier tenets. As Lester (2021) suggests, perhaps post qualitative could be considered relational rather than oppositional, however, I suggest that there may be more work to do in exploring this idea in greater depth in PQI.

It is not the purpose nor intention to challenge St. Pierre's proposition in PQI to provide a new or different interpretation. I have a deep respect for St. Pierre's work in PQI and it is from this respect that I have wanted to uphold the integrity of PQI through which St. Pierre introduced and continues to advocate for. However, I have been critiqued for too narrowly exploring the field of PQI and failing to read other perspectives. I admit that I find this work challenging for two reasons, (i) I do not claim to be a post qualitative inquirer and (ii) see a field that is rife with varied re-interpretations; often differing and in conflict with what St. Pierre has established.

St. Pierre (2011) did not define PQI, however, principles and tenets have been introduced to guide inquiries and engagements. These aspects of PQI have been explored so far in this Turn as (i) through adopting poststructuralist thinking, (ii) PQI providing an avenue for theorising research into existence, and (iii) undertaking inquiry – not research – without a methodology, a methodological approach or methods. In doing this work, PQI has changed the way research is considered and known. Equally so, it has demonstrated that it is possible to work outside the boundaries of qualitative and quantitative research. And, echoing Lester's (2021) position, this is incredibly important work.

Problematising post qualitative inquiry

The possibilities and potentialities for PQI are endless and St. Pierre's thoughtful and considered work over the many decades is evidence of this. St. Pierre (2023, p. 21; emphasis in original) stated,

> I've written a great deal about the incommensurability between what I've called "conventional humanist qualitative methodology" and poststructuralism. But, as a doctoral student, I studied them at the same time, *although separately*.

They never came together until I wrote my dissertation – it was in writing that their incommensurability became apparent.

I empathise with these feelings of incommensurability. I started the research in this book – my PhD research – thinking I would conduct a PQI. I soon realised the incommensurability between what I wanted to do in adopting methods adapted from qualitative ethnography with posthuman theories and PQI. The more I read about PQI, the more the tension and conflict grew. What I wanted to do and what post qualitative described by St. Pierre who coined the term and introduced the field, seemed incommensurable.

I am not the first (nor likely the last) to problematise PQI. As is inherent in researchers who seek to challenge conventional research practices, there is the tendency to look at something from every angle, turn it upside down, inside out, and back again. As interrogators who are researchers or inquirers, there is no stationary, fixed point from which to start. Murris (2020a) describes the dynamic nature of PQI – as the philosopher Heraclitus did – as being "in flux". This philosophy, which is also a science, was brought together through Barad's (2007) work. Barad (2007) provided a bridge to bring the wonder of the micro-physical world into and through the social sciences. Barad (2007) made clear and simple the fact that – on a micro-scale – you cannot truly repeat anything and get the same result. Similarly, as Heraclitus (cited in Murris, 2020a) stated that you cannot "step in the same river twice". I then question, and perhaps challenge, St. Pierre (2023) in how they find the "functionary of the method so dreadfully boring" (p. 23). If every step in the river is a "new" and different river, how can method repeated ever be the same?

I embrace method. I enjoy knowing how to approach something. To learn its history. To see how an approach to something can change in the subtlest and minutest of ways. I enjoy process and appreciate it when processes, approaches and methodologies are clear, simple and accessible. I do not find it boring; it can be supportive and enriching. I deeply appreciate processes that make our lives simple. And I see process everywhere! In First Nations knowledges and how they methodologically shared stories and rituals over thousands and thousands of years; in nature where process is in the cycles of day and night, the seasons, life and death; in the way stars are formed, and unformed. I considered that perhaps it is the way we approach methods and not the methods themselves that are the problem.

In my PhD research, I wanted to learn about and understand the research culture that had been laid down in the academy. I wanted to know what drove research processes that had formed the field we broadly accept as research. Qualitative research methods did make sense to me. They had a clear process and I saw the value in talking with people to understand their thinking and practice (ethnographic interviews) and have always used observation to learn (ethnographic lesson observation). However, I did want to problematise the focus on humans at the silencing of nature, and the relevance of replication, objectivity and binaries. I also wanted to disrupt the harm experienced in silencing First Nations knowledges and voices. I saw the posthuman as a theoretical way to work with these ideas and apply them to qualitative methods (I will explain in greater detail how I approached this later in this Turn).

Midway through my candidature, I could not resolve the conflict I was experiencing with a methodology. I played with the idea of nestling my research in a liminal space between qualitative and PQI; however, it seemed like an oxymoron – too juxtaposed to make ontological sense. My supervisors encouraged me to dig deep and consider how to approach my research in a way that was commensurate with posthuman theory with qualitative research practices. It finally landed. I did not have to undertake a PQI. I did not have to change my onto-epistemology (Barad, 2007), to fit into PQI. Indeed, I needed to take a different kind of learning from St. Pierre (2023) and consider how to think about research differently that worked with my onto-epistemology. As St. Pierre (2014b, p. 15) reminds us,

> we must remember that we invented qualitative methodology as an interpretive research methodology to counter positivist social science almost 30 years ago. We invented it. We made it up. It's not sacred. The sky won't fall if we just put it aside and try something different.

In the last few years, other approaches to undertaking research that is not qualitative nor post qualitative have begun to emerge. Such research troubles some of the values and principles established in PQI that create a barrier to its use (Aagaard, 2022; Ellingson & Sotirin, 2020). For example, Ellingson and Sotirin (2020), trouble the post qualitative position that seeks to do away with data (Denzin, 2013) and offer a "generative alternative" that adopts "intersectional feminist and other critical, materialist theorizing" to "articulate a methodological practice that incorporates making, assembling, and becoming data, along with ethical commitments to pragmatism, compassion, and joy" (Ellingson & Sotirin, 2020, p. 817). They call this methodological practice "data engagement" (Ellingson & Sotirin,

2020). Another example is Aagaard (2022), in their aptly titled "Troubling the troublemakers", who poses several challenges and conflicts presented by PQI. Most predominantly, Aagaard (2022) puts forth three arguments that PQI risks in "theory-centrism, researcher deletion, and meta-reflexivity" (Aagaard, 2022, p. 311). In taking on these risks, Aagaard (2022) warns that many post qualitative values may end up as "constraining dogmas" (p. 321). In doing so, post qualitative may restrict the very research it was designed to support. Another emergence is performative research (for example, Bolt, 2016; Menning & Murris, 2023; Østern et al., 2023), a creative approach that caters for artistic as well as post qualitative research where "research is understood as creation" (Østern et al., 2023, p. 273). Springgay and Truman (2018) suggest not doing away with methods and hold their position as qualitative researchers who trouble methods by approaching them through propositions, speculations and experiments. They argue that it is the "logic of procedure and extraction that needs undoing" (p. 204), "rather than a refusal of methods" (p. 203).

Others from the PQI field have too acknowledged the problematic nature of working within a space that is "inspired by the ontologies of difference and relationality and a methodological thinking that appears fluid, unstable and with uncertain and indeterminate dimensions" (Murris, 2020a, p. 11). These tenets make PQI slippery and difficult to define; which can be challenging when working in the field of "research" which was founded on process and systematic thinking.

Other challenges have recognised the dissonance within PQI. For example, Murris (2020a) in the opening chapter of the first book in this series shares the further difficulty facing PQI where

St. Pierre (2014a), who coined PQI, establishes boundaries and frameworks that are in conflict with how other researchers have interpreted and put PQI to work in their own research/inquiry. St. Pierre (2021b) clearly articulates that PQI *"is not a methodology at all"* (p. 63; emphasis in original). Murris (2020a) identifies this as a point of tension, while Young et al. (2021) seek to disrupt the notion that PQI resists qualitative methodologies by exploring a "disruption" titled, "Disruption: post-qualitative research rejects qualitative methodologies" (p. 313), which also troubles St. Pierre's (2023) call that post qualitative is not research, but inquiry.

MacLure, who noticeably resists entering into discussions about what PQI is in their own research, seemingly prefers to teach through example and putting PQI to work (for example, MacLure, 2013, 2024a, 2024b). MacLure's inquiries provide great examples of PQI in practice through leaning heavily into thinking with poststructural theory as a first point of engagement and a lack of methodologies and methods. MacLure (2023), seemingly somewhat reluctantly by not being "invested in the contemporary debate for or against methods" (p. 217), shares a discussion with Candace Kuby on the topic. MacLure (2023) provides insight into their thoughts about methods – acknowledging that it matters not whether they are called methods or protocols – and states "you need methods, but they need to be bespoke methods … you do need to have some sense of method, even if it unfolds from – must unfold from – ongoing, immanent immersion in the field" (p. 218). In contrast, St. Pierre (2021a) adopts a poststructuralist position to put forth the argument that method and thought or thinking do not align – quoting Deleuze's anti-method and Derrida's contribution (as cited in St. Pierre, 2021a), "a thinker with a method has already decided how to proceed and is simply a functionary of the method, not a thinker" (p. 4).

While I acknowledge how the research informing this book aligns with post qualitative thinking – such as thinking with posthuman concepts first, never considering myself as the researcher outside or separate from the data, adopting creative approaches and not delineating between data collection and analysis – I found unresolvable incongruencies between the research underpinning this book and PQI as proposed by St. Pierre, who coined the term. These misalignments involve where I have adopted (i) a lack of poststructuralist theoretical underpinning, (ii) the use of qualitative methods and methodologies, and (iii) the adherence to a strict research process. Where the research in this book takes a turn away from conventional qualitative research and into the transqualitative is in the conceptualisation of the methods through the posthuman lens, and also in the reconceptualisation of data collection and analysis processes not as separate entities, but rather, they are described in this book as an intrinsically, intra-acting and entangled process called diffractive data entanglements.

Unlike PQI, transqualitative inquiry does not prescribe that there is no research design or research process (St. Pierre, 2019, p. 9). Transqualitative inquiry embraces research conventions that have existed before and pushes them in theoretical ways that more aptly speak to current global situations and enact critical and creative responses. Transqualitative research is sympathetic to research structures that include naming a methodology. St. Pierre (2019, p. 10; emphasis in original) strongly advises that,

there can be no post-qualitative research methodology or research methods, no post-qualitative research designs, no post-qualitative research practices, no post-qualitative data or methods of data collection or methods of data analysis, no representations of a stable, sensory "lived" world, no post-qualitative findings, no post-qualitative research report format because, again, post-qualitative inquiry never is, it never stabilizes.

As the research documented in this book involves ethnographic methods, posthumanly realised, and conventional research processes, it emerged that it [the research informing this book] could not be appropriately described as post qualitative. The research was seeded from a question and engaged deeply with methods such as conversations and lesson participations to think with posthuman theory. From this place, transqualitative research emerged as a methodological approach that embraced these qualitative methods, through a posthuman lens, which enabled creative and innovative ways to approach data, and accessible research approaches that I still considered research.

Transqualitative research as reconciliation

The void that I felt when I could not reconcile my methodology in alignment with my theoretical position and my methods and methodological approach is in itself a diffractive metaphor. What is the potentiality of the space left when qualitative and post qualitative methodologies are diffracted through each other? As stated by Mitchell (2017), "like waves interfering with each other in the ocean, a new in-between space is opened up, like new patterns that are formed in/through the water" (p. 174). This space calls the transqualitative into business as a methodology

that accepts the limitations of conventional qualitative paradigms while not fighting against them to provide a reconciliation
of the methodology void.

Transqualitative research enables the diffraction of traditional
humanistic qualitative approaches, such as ethnography, with/
in/through posthuman approaches such as diffraction to intra-act
and generate "new" and different ways of doing data. Transqualitative research offers a methodology that accepts the tensions to
explore the possibilities when qualitative ethnographic methodologies are pushed through posthuman thinking such as through
the diffractive lens.

Transqualitative research

Transqualitative inquiry offers a different direction for research
that pushes beyond the limitations and structures of more conventional quantitative and qualitative research frameworks, even
the more emergent, PQI. For some, another methodology may
seem unnecessary and a step away from what they consider
meaningful, qualifiable, perhaps even quantifiable and valid
research. This is understandable given the seeming swift movement from almost solely quantitative, to qualitative, to PQI in
research over the last 50 years. However, while in many ways I
appreciate the opportunity to slow things down, there are many
instances where I think we do not have the time to act slowly:
nature's signal is perhaps one of the most urgent of these signals.
The immediacy of changes that are being experienced globally
across social, economic, political, and environmental domains
are indeed unprecedented and this call is what I respond to with
research that challenges and re-challenges what is considered
valid research.

To fit the research underpinning this book into a space between qualitative and PQI, I was troubled that the research was not considered wholly post qualitative nor qualitative and attempted to justify a new mixed methodology or the liminal space in-between. However, given the ethico-onto-epistemological position of my study and the way I was approaching the research, the lack of methodology that aligned theory and practice became increasingly problematic. Qualitative and post qualitative methodologies both had well-established paradigms and structures that did not align with this research – both theoretically and methodologically. As neither approach was fitting, a mixed methodology (qualitative/post qualitative) was inappropriate. Something "new" and different was required. I proposed the transqualitative methodology to provide a fluid research space without rigidity or strict parameters to enable creative, innovative and expansive thinking to be explored. Drawn from the Latin, where trans- is the prefix to mean *across, beyond* and *through*, transqualitative inquiry espouses pushing beyond the current research methodologies and approaches to enable research that follows a methodological process but also allows for the creative, out-of-the-box thinking, research and knowledge generation that is needed in this spacetime.

Transqualitative inquiry does not specify a theoretical alignment such as post qualitative (which must be tied to poststructuralist thinking). The openness in theoretical positioning is an integral aspect of progressing beyond current ways of thinking and exploring truly different realms and possibilities in research. Transqualitative inquiry is applied through the lens of education in this book, however, it is hoped that other researchers find this methodology useful to explore research in education and other disciplines to enact transdisciplinary approaches.

Transqualitative as a research methodology has been explained previously by Rousell and Cutter-Mackenzie-Knowles (2018) in describing a "parental milieu" (p. 93). It is clearly defined in this instance but does invite keeping "the concept of the parental milieu open to further trans-qualitative movements and potentials" (p. 93). In addition, in a synchronous movement while I was writing about transqualitative here, Rousell (2021) explored transqualitative as "trans-qualitative" under PQI to describe a plethora of concepts such as milieu, affect, ecology, patterns, engagement, experience, movement, learning environment, theory of learning, passage, process, and transformation, and espouses that "the term evokes a diverse genealogy of engagements with the prefix 'trans'" (p. x). So, while the research in this book does not work with the concept of the parental milieu nor use transqualitative as a describing word, it accepts the challenge to expand transqualitative research to realise its potential as a methodology, which is inclusive of many more concepts, theories and creative research approaches.

Transqualitative research is *transitional*. Research methodologies should not be considered a static or fixed instructional approach to exploring and understanding data. Time does not stand still in a space of nothingness. Similarly, the planet is changing and evolving every moment which requires a responsiveness to the way research is undertaken. Methodologies need to be nimble to accommodate the unknown future while allowing for institutionalised structures and procedures to be adhered to.

Transqualitative research is *transdisciplinary*. As described earlier, the prefix trans- also means *across*. Transqualitative inquiry is established to enact research that combines different ontological and epistemological positions that stem from different disciplinary fields. Disciplines generally have a research methodology that they adhere to, the well-worn paths and traditions of the academy. Transqualitative inquiry seeks to trouble these limiting factors by proposing that transdisciplinary approaches are much needed to allow for creative ways of exploring research questions and problems.

Conclusion

This Turn provided an overview of qualitative research to highlight its human-centred focus and the tensions created when adopting posthuman theorising. Through the tension, transqualitative inquiry emerged to enable posthuman research that also leaned into qualitative practices, such as ethnography. Transqualitative research is a move beyond conventional qualitative research to enable transdisciplinary research that entangles posthuman theorising with educational practice while acknowledging other theories could be suitably used instead. Transqualitative research is also transitional to enable research to dwell in the dynamism and flux of research practices. I justified against adopting a PQI based on a misalignment between my research theory and practice using methods. Transqualitative research proposed a reconciliation that enables creative and different ways of doing data to be theoretically and methodologically robust. In addition, transqualitative inquiry posed as reconciliation, in this instance, between posthuman thinking and qualitative methods and methodological approaches.

Note

1 Worth noting is that the poststructural theorists described in Foucault, Derrida, Deleuze, and Guattari are from a discrete point in

spacetime and represent a sliver of the population, that being of male, white, middle-class privileging where the major works were produced in the narrow timeframe of the 1960s–1990s and from France. Moreover, it is important to consider if these texts are read in their original French language or read through the translation. This impacts the way the texts are interpreted and understood which can greatly impact the way they are then embodied and lived (Lasczik, personal communication, 7 December 2021). The question arises, what then for research that problematises human-centred research, not informed by poststructualist thinking? I was troubled by this question in justifying a research process that was not informed by poststructuralism but posthumanism specifically.

References

Aagaard, J. (2022). Troubling the troublemakers: Three challenges to post-qualitative inquiry. *International Review of Qualitative Research*, *15*(3), 311–325. https://doi.org/10.1177/19408447211052668

Barad, K. (2007). *Meeting the universe halfway: Quantum physics and the entanglement of matter and meaning*. Duke University Press.

Barad, K. (2014). Diffracting diffraction: Cutting together-apart. *Parallax*, *20*(3), 168–187. https://doi.org/10.1080/13534645.2014.927623

Barad, K., & Gandorfer, D. (2021). Political desirings: Yearnings for mattering (,) differently. *Theory & Event*, *24*(1), 14–66.

Bolt, B. (2016). Artistic research: A performative paradigm. *Parse Journal*, *3*(1), 129–142.

Davies, B. (2021). *Entanglement in the world's becoming and the doing of new materialist inquiry*. Taylor & Francis Group.

Denzin, N. K. (2013). The death of data? *Cultural Studies ↔ Critical Methodologies*, *13*(4), 353–356. https://doi.org/10.1177/1532708613487882

Denzin, N. K., & Lincoln, Y. S. (2008). *The landscape of qualitative research* (3rd ed.). Sage Publications.

Ellingson, L. L., & Sotirin, P. (2020). Data engagement: A critical materialist framework for making data in qualitative research. *Qualitative Inquiry*, *26*(7), 817–826. https://doi.org/10.1177/1077800419846639

Gullion, J. S. (2018). *Diffractive ethnography*. Routledge. https://doi.org/10.4324/9781351044998

Jackson, A. Y., & Mazzei, L. (2011). *Thinking with theory in qualitative research: Viewing data across multiple perspectives*. Routledge.

Koro, M. (2021). Post-qualitative projects: Exhilarating and Popular "fashion"? *Qualitative Inquiry, 27*(2), 185–191. https://doi.org/10.1177/1077800420932600

Koro-Ljungberg, M. (2015). *Reconceptualizing qualitative research: Methodologies without methodology*. Sage Publications.

Koro-Ljungberg, M., MacLure, M., & Ulmer, J. (2018). D… a… t… a…, data++, data, and some problematics. *The Sage Handbook of Qualitative Research, 5*, 462–484.

Lester, J. N. (2021). Relational engagements with post-qualitative inquiry: There are no blank pages. *Qualitative Inquiry, 27*(2), 219–222. https://doi.org/10.1177/1077800420931133

Loveless, N. (2019). *How to make art at the end of the world: A manifesto for research-creation*. Duke University Press.

MacLure, M. (2013). Researching without representation? Language and materiality in post-qualitative methodology. *International Journal of Qualitative Studies in Education, 26*(6), 658–667. https://doi.org/10.1080/09518398.2013.788755

MacLure, M. (2023). Ambulant methods and rebel becomings: Re-animating language in post-qualitative inquiry. *Qualitative Inquiry, 29*(1), 212–222.

MacLure, M. (2024a). Resistance, desistance: Bad girls of post-qualitative inquiry. *International Journal of Qualitative Studies in Education, 37*(3), 631–641. https://doi.org/10.1080/09518398.2022.2127025

MacLure, M. (2024b). "Something comes through or it doesn't": Intensive reading in post-qualitative inquiry. *International Journal of Qualitative Studies in Education*, 1–8. https://doi.org/10.1080/09518398.2024.2342696

Mazzei, L. A. (2021). Postqualitative inquiry: Or the necessity of theory. *Qualitative Inquiry, 27*(2), 198–200. https://doi.org/10.1177/1077800420932607

Menning, S. F., & Murris, K. (2023). Reconfiguring the use of video in qualitative research through practices of filmmaking: A post-qualitative cinematic analysis. *Qualitative Research, 24*(4), 1000–1020.

Mitchell, V. A. (2017). Diffracting reflection: A move beyond reflective practice. *Education as Change, 21*, 165–186. http://www.scielo.org.za/scielo. php?script=sci_arttext&pid=S1947-94172017000200010&nrm=iso

Murris, K. (2020a). Introduction: Making kin: Postqualitative, new materialist and critical posthumanist research. In *Navigating the postqualitative, new materialist and critical posthumanist terrain across disciplines* (pp. 1–21). Routledge.

Murris, K. (2020b). Posthuman child and the diffractive teacher: Decolonizing the nature/culture binary. In A. Cutter-Mackenzie, K. Malone, & E. Barratt Hacking (Eds.), *Research handbook on childhoodnature: Assemblages of childhood and nature research* (pp. 1–25). Springer International Publishing. https://doi.org/10.1007/978-3-319-51949-4_7-2

Murris, K., & Kohan, W. (2021). Troubling troubled school time: Posthuman multiple temporalities. *International Journal of Qualitative Studies in Education, 34*(7), 581–597.

Musante, K., & DeWalt, B. R. (2010). *Participant observation: A guide for fieldworkers*. Rowman Altamira.

Nelson, C., Treichler, P. A., & Grossberg, L. (1992). Cultural studies: An introduction. In L. Grossberg, C. Nelson, & P. A. Treichler (Eds.), *Cultural studies* (Vol. 1, pp. 1–16). Routledge.

Østern, T. P., Jusslin, S., Nødtvedt Knudsen, K., Maapalo, P., & Bjørkøy, I. (2023). A performative paradigm for post-qualitative inquiry. *Qualitative Research, 23*(2), 272–289.

Rousell, D. (2021). *Immersive cartography and post-qualitative inquiry: A speculative adventure in research-creation*. Routledge.

Rousell, D., & Cutter-Mackenzie-Knowles, A. (2018). The parental milieu: Biosocial connections with nonhuman animals, technologies, and the earth. *The Journal of Environmental Education*, 1–16. https://doi.org/10.1080/00958964.2018.1509833

Silverman, D. (2016). *Qualitative research* (4th ed.). Sage Publications.

Snaza, N., Sonu, D., Truman, S. E., & Zaliwska, Z. (2016). *Pedagogical matters: New materialisms and curriculum studies*. Peter Lang Publishing, Inc.

Springgay, S., & Truman, S. E. (2018). On the need for methods beyond proceduralism: Speculative middles, (in)tensions, and responseability in research. *Qualitative Inquiry, 24*(3), 203–214. https://doi.org/10.1177/1077800417704464

St. Pierre, E. A. (2011). Post-qualitative research: The critique and the coming after. *The Sage Handbook of Qualitative Research, 4,* 611–626.

St. Pierre, E. A. (2013). The appearance of data. *Cultural Studies ↔ Critical Methodologies, 13*(4), 223–227. https://doi.org/10.1177/1532708613487862

St. Pierre, E. A. (2014a). A brief and personal history of post-qualitative research: Toward "post inquiry". *Journal of Curriculum Theorizing, 30*(2). https://journal.jctonline.org/index.php/jct/article/view/521

St. Pierre, E. A. (2014b). *Post-qualitative inquiry.* Australian Association of Research in Education (AARE).

St. Pierre, E. A. (2018). Writing post-qualitative inquiry. *Qualitative Inquiry, 24*(9), 603–608. https://doi.org/10.1177/1077800417734567

St. Pierre, E. A. (2019). Post-qualitative inquiry in an ontology of immanence. *Qualitative Inquiry, 25*(1), 3–16.

St. Pierre, E. A. (2021a). Post-qualitative inquiry, the refusal of method, and the risk of the new. *Qualitative Inquiry, 27*(1), 3–9.

St. Pierre, E. A. (2021b). Why post-qualitative inquiry? *Qualitative Inquiry, 27*(2), 163–166. https://doi.org/10.1177/1077800420931142

St. Pierre, E. A. (2023). Poststructuralism and post-qualitative inquiry: What can and must be thought. *Qualitative Inquiry, 29*(1), 20–32.

St. Pierre, E. A. (2024). Reading for post-qualitative inquiry. *International Journal of Qualitative Studies in Education, 37*(2), 1–11.

Sweet, J. D., Nurminen, E., & Koro-Ljungberg, M. (2020). Becoming research with shadow work: Combining artful inquiry with research-creation. *Qualitative Inquiry, 26*(3–4), 388–399.

Taguchi, H. L., & St.Pierre, E. A. (2017). Using concept as method in educational and social science inquiry. *Qualitative Inquiry, 23*(9), 643–648. https://doi.org/10.1177/1077800417732634

Taylor, C. A. (2020). Knowledge matters: Five propositions concerning the reconceptualisation of knowledge in feminist new materialist, posthumanist and postqualitative approaches. *Navigating the post-qualitative, new materialist and critical posthumanist terrain across disciplines* (pp. 22–42). Routledge.

Tesar, M. (2021). Some thoughts concerning post-qualitative methodologies. *Qualitative Inquiry, 27*(2), 223–227. https://doi.org/10.1177/1077800420931141

Truman, S. E. (2023). Undisciplined: Research-creation and what it may offer (traditional) qualitative research methods. *Qualitative Inquiry, 29*(1), 95–104. https://doi.org/10.1177/10778004221098380

Verlie, B. (2019). *Affective entanglements: Learning to live-with climate change* (Doctoral dissertation, PhD thesis, Monash). https://bridges.monash.edu/articles/thesis/Affective_entanglements_Learning_to_live-with_climate_change/7901675?file=14725727

Young, T., Crinall, S., & Malone, K. (2021). Disruptions of post-qualitative education research: Tensions and openings. *Qualitative Inquiry, 28*(3–4), 312–321.

5

THE FIFTH TURN

Diffractive ethnography as a posthuman qualitative tension in environmental education research

Introduction

This Turn explores the methodological approach of diffractive ethnography that has provided an emerging field founded in qualitative research traditions to guide the inquiry presented in this book. As described in the Fourth Turn, this book explores early school years teachers' nature perceptions and pedagogies through a posthuman theoretical position and adopts a transqualitative methodology. Transqualitative research emerged through the work in this book and enacts qualitative methodological approaches and methods through a posthuman framework. The research underpinning this book embraces ethnography through a diffractive lens. Ethnography accommodates diversity across methods, applications, and longevity (Gobo & Marciniak, 2016) and is used diffractively here to disrupt the human focus.

Ethnography

At its most basic, ethnography is the study of people and their social interactions and cultures (Gobo & Marciniak, 2016; Gullion,

DOI: 10.4324/9781032703473-5

2018). Ethnography has been interpreted in myriad ways, particularly across disciplines but within disciplines too (Denzin & Lincoln, 2011; Hammersley & Atkinson, 2007). Hammersley (2018) problematises these reinterpretations claiming they may impact the continuance of ethnography as a field of research. These concerns were shared by Ingold (2014) who attests that ethnography is literally *"writing about people"* (p. 385; emphasis in original).

Theoretically, ethnography was originally proposed from a default "modern realist" perspective (Denzin & Lincoln, 2011; Gullion, 2018). More recently ethnography has exploded into a plethora of varied theoretical spaces. Hammersley (2018) uses the term "qualifying adjectives" (p. 5) to describe more than 40 areas that ethnography has moved into beyond the "modern realist" position, for example, feminist ethnography (Visweswaran, 1994), autoethnography (Ellis et al., 2011), sensory ethnography and visual ethnography (Pink, 2012, 2015), performance ethnography (Denzin & Lincoln, 2011), and multispecies ethnography (Lloro-Bidart, 2018; Pacini-Ketchabaw et al., 2016). While Hammersley (2018) troubles this deviation, primarily due to their lack of purpose in contributing to the knowledge of and for the social world (Hammersley, 2018, p. 12). Notwithstanding Hammersley's (2018) perspective, it is unlikely the trend in ethnographic variances is going to slow given the emergence and uptake in recent times and given their longevity in research; such as the 40+ year history of autoethnography (see Holman Jones et al., 2013). This precedent has enabled further variances such as the diffractive ethnography underpinning this book. Diffractive ethnography brings attention to nonhuman others and as such, challenges the conventional and understandably humanistic lean of ethnography. In doing so, diffractive ethnography stays in the tension of applying posthuman thinking to an ethnographic (human-centred) study.

Diffraction

> If the goal is to think the social and the natural together, to take account of how both factors matter (not simply to recognise that they both do matter), then we need a method for theorizing the relationship between "the natural" and "the social" together.... What is needed is a diffraction apparatus to study these entanglements.... A diffractive methodology provides a way of attending to entanglements in reading important insights and approaches through one another.
>
> Barad (2007, p. 30)

As is described in this opening quote by Barad (2007), diffraction is a way of knowing the world from within and as a part of it, while providing a theoretical position to allow for "new" knowledge to emerge. It offers a way of understanding the relationality of bodies, objects, and spaces through their materiality. The concept of diffraction originates from classical physics to describe the pattern of overlapping waveforms that disturb a field. As a model for understanding inter- and intra-actions, diffraction describes the way separateness between entities is non-existent; that is, there is no distinct outside (Barad, 2007). Barad (2014) explains that "a diffraction pattern does not map where differences appear but ... where the effects of difference occur" (p. 172). Diffraction provides an astute example of the entanglement of matter. It describes matter as a dynamic relationality that challenges particularism. In doing so, the intra-active and emergent nature of beings and phenomena is brought into focus (Barad, 2007).

The use of diffraction as an approach to considering research was first introduced by Haraway (1992) and then elaborated on by Barad (2007). Barad (2007) described diffraction as "a way of understanding the world from within and as part of it" (p. 88). Diffractive approaches to research consider that the researcher is always already a fundamental part of the research becoming. In adopting this view, the researcher is entangled with the research in a way that does not require justification, accountability or reflexivity where "securing objectivity in research is not about disentangling and disengaging the subject from the object (as in much research), but 'taking responsibility for one's entanglements' (Barad, 2007, p. 453, footnote 1)" (Murris, 2020, p. 8). Throughout my research, I do not attempt to position myself as an "outsider researcher" with some form of objective position but accept my position as part of the research becoming.

Following on, Barad (2007) introduced diffractive methodology acknowledging Haraway's work in its conceptualisation. Barad (2011, p. 10) explains a diffractive methodology as,

...a practice of simultaneously attending to important differences among practices and their specific entanglements by reading respective insights through one another in a way that does not build in foundational distinctions and separations before the analysis gets off the ground, and that is responsive to our intra-active engagements with our subject matter, including attending to what gets excluded and how it matters.

Here, Barad brings attention to the significance and necessity of considering the materiality and relationality of matter in a diffractive methodology. Using diffraction through an ethnographic process as the methodological approach adopted in this study decentres the human, and highlights the materiality of all matter. Relationality is centred instead where the human "subject no longer lies at the centre of all meaning-making, and is no longer, collectively, the sole agent commanding the worlding of the world" (Davies, 2021, p. 140). Relationality adopts a relational ontology where human privileging is put on notice such that nonhuman agency too has attention as demonstrated in this book. This point is discussed in greater detail below (see principle four, *the choice of data matters*).

The influence of Barad's seminal work is demonstrated by the huge upwelling of social science research based in diffractive thinking and practice over the last decade (for example see Barad, 2014; Bozalek & Zembylas, 2017; Elfström Pettersson, 2017; Kaiser & Thiele, 2014; Lennon, 2017; Lenz Taguchi, 2012; Malone, 2020; Murris, 2017, 2020; Murris & Haynes, 2018; Sehgal, 2014; Taylor, 2013; Taylor & Blaise, 2014; van der Tuin, 2011). Moreover, it has become increasingly popular in educational research (Bozalek, 2017; Larson & Phillips, 2013; Malone, 2020; Murris & Haynes, 2018; Taylor, 2013) including the introduction of diffractive pedagogies (Moxnes & Osgood, 2019), which establish alternative ways of understanding a range of educational practices and exploring future possibilities.

In further justification of diffraction, Murris (2020 , p. 21) stresses how this methodological approach could offer an education revolution since,

> ...diffraction helps materialize important new insights for post-human schooling. It disrupts the idea of humanist schooling that knowledge acquisition is mediated by the more expert and knowledgeable other; schooling as a linear journey from child to the more "fully-human" adult. Importantly, *the diffractive teacher can be human, nonhuman or more-than-human,*

contributing to a reconfiguration of the world in all its materiality – a process of "worlding." Importantly, this process is always *relational,* not *individual.*

Here, Murris decentres the human and enacts the agency of matter including the role of the nonhuman, for example, the role of nonhuman nature in being an educator and teacher for children and students. In doing so, diffraction opens up possibilities for complex and multiple perspectives, perceptions and conceptualisation to be explored.

Diffractive ethnography

In a diffractive ethnography, the researcher is a presence, and active force….

Gullion (2018, p. 122)

In this opening quote, Gullion highlights that, as noted previously in this Turn, diffraction decentres the human while also acknowledging that objectivity is not about taking the subject out of the research (Murris, 2020), but taking responsibility for our entanglements (Barad, 2007). The researcher is indeed "a presence, and active force" that cannot be removed or silenced from the research they are undertaking. In adopting a diffractive ethnographic approach, it is hoped that "the missing voice in that discourse – the silence of matter" (Gullion, 2018, p. 1) is mediated to consider more richly all the material-discursive discourses that are taking place simultaneously in the research. For example, lesson observations are reframed as lesson participations that consider not only the human participants[i] but also the nonhuman participants and the nonrepresentational aspects such as the social, cultural and other material-discursive forces. This is a necessary practice of reconceptualisation in posthumanist research (Murris, 2020), and there is more on this to come in the discussion on methods.

Gullion (2018) draws on theories of assemblages, difference and becoming as described by Deleuze and Guattari's network actor theory and associations from Latour, and entanglements as offered by Barad. Gullion (2018) defines that the concepts of assemblage, entanglement and mangle will be used interchangeably. With respect to the scientific theories of quantum mechanics underpinning the research informing this book, I focus here solely on entanglements (as was justified in the Fourth Turn). The purpose is to dive deeply into the complexity of "entanglements" as a key concept in Barad's work without reducing or negating its meaning by interchanging it with other terms such as enmeshed, assemblage or network. The concept of entanglements in diffractive ethnography aligns with the posthuman theoretical perspective. In particular, diffractive ethnography encourages the questioning of the individual *self* where "existence is not an individual affair. Individuals do not preexist their interactions; rather, individuals emerge through and as part of their entangled intra-relating" (Barad, 2007, p. xi).

Diffractive ethnography adopts conventional ethnographic methods such as observations and interviews and transforms them through a shift in thinking – from human-centric to posthuman. Gullion (2018) describes this change in thinking from one that focuses on people to one that seeks to explore the nature and relations of entanglements through phenomena; both human and nonhuman alike. Diffractive ethnography is a posthuman research practice where the focus on phenomena enables the study of myriad relations and their complexities (Ulmer, 2017).

Diffractive ethnography requires a commitment to dwell in the tension of adopting humanistic approaches through posthuman thinking. Diffractive ethnography requires researchers to focus on the entanglements of materiality and relationality of matter and on the way data make themselves known. In doing so, diffractive ethnography encourages the researcher to explore "how things come to matter in the ways they do" (Davies et al., 2013, p. 680). Rather than conventional "cause-and-effect models", diffractive ethnography brings the flows and patterns of diffraction to attention (Davies et al., 2013; Gullion, 2018; Mitchell, 2017; Ulmer, 2017). This provides a further tension and challenge as diffractive ethnography grapples with data representation in what is frequently described as a nonrepresentational paradigm (Davies et al., 2013; Gullion, 2018; Mitchell, 2017).

In the following passage, Gullion (2018 , p. 156) elucidates the nature of diffractive ethnography, including reconceptualising society and ethical responsibility,

In the ontological turn, we can reconceptualize the social world as an enchanted place of entanglements … of dynamics and movements that include humans and others. Thinking with entanglements … incites issues of ethics and justice…. Harm to entities that are entangled with us harms us directly as well…. When an object is entangled, distance is erased. Boundaries between entities are enactments; indeed, to bound an entity itself is an enactment, and we are responsible for those enactments. I should be clear, however, that these are discursive; there are no "real" boundaries between us. Speaking in quantum terms, there is no separation between entities. We are entangled.

This passage explains the entangled nature of diffractive ethnography and how it offers a methodological approach to research that constantly shifts the angle of perception of ways of knowing and being in the world. Diffractive ethnography encourages researchers to look beyond just the human figure and take into consideration the agency of all matter; including the material-discursive forces that are unseen but make an impact. Rather than seeing human bodies each as an isolated self, diffractive ethnography aligns with an ethico-onto-epistemology where human bodies are perceived as an entanglement. Diffractive ethnography provides a methodological approach where individuals "emerge through and as part of their entangled intra-relating" (Barad, 2007, p. ix). As such, "the boundaries that we perceive are not real" (Gullion, 2018, p. 112).

Diffractive ethnography adopts ethnographic methods and thinking through a posthuman, diffractive lens – the ReTurning Learning theoretical framework – as a series of diffraction gratings (see Figure 4.1 in the Fourth Turn). Diffractive ethnography tends to the uncertain and indeterminate, by entangling the posthumanist theorising of diffraction with the qualitative tradition of ethnography to reconceptualise the complexity, intra-activity and agency of data in research.

The e/mergence of diffraction and ethnography in a diffractive ethnographic approach was proposed theoretically by Gullion (2018) and adopted in this study, was still in its infancy at the time of writing this book. While I acknowledge the efforts and early hints of diffractive ethnography that were proposed by earlier researchers such as Schneider (2002), and Mellander and Wiszmeg (2016), the grounding of diffractive ethnography has been largely evident through the book, *Diffractive Ethnography,* released by Gullion in

2018. A Google Scholar search for "Diffractive Ethnography" any-time until 2017 returned 31 hits. This compares with the 42 hits since the book's release in 2018. Therefore, while this is an emerging and burgeoning area of research, there are still few empirical studies that have utilised diffractive ethnography in practice. This PhD study is therefore significant in grounding the practical implementation of diffractive ethnography. Furthermore, this research makes a unique and novel contribution to the field of environmental education as this approach to research has not been utilised in this educational area to date.

I now move to explain the data entanglement process by first outlining the principles that inform that practice/process.

Doing data: the principles underpinning the practice

In consideration of how to work with the data with the nonrepresentational aspects of posthuman research, I decided to present the data through a series of diffractive data entanglements (for a more fulsome explanation, see the Fourth Turn). Diffractive data entanglements describe the combined data collection and analysis processes as an iterative event where lines are blurred and data is generated as part of the research. Data "that speaks" was given voice in response to the research question (Murris & Haynes, 2018; Taylor, 2013) and data emerged and was generated through a theory-making practice (Gullion, 2018). Jackson and Mazzei (2011, p. 11) describe this as the process of "flattening and folding" that recoils from interpretivist notions of data analysis and "embraces the mutually constitutive nature" of diffraction.

Data analysis with its interpretation, coding and categorising of data, focuses on sameness and repetitive themes. This differs from diffraction which is "attentive to how differences get made and what the effects of these differences are" (Bozalek & Zembylas, 2017, p. 2). Diffraction considers the "relations of difference and how they matter" (Barad, 2007, p. 71).

In this book, the data guided and led the process of exploration; data resisted interpretation, being condensed and packaged up neatly. The data were given the opportunity to expand thought and inquiry into (yet) unknown and potentially uncomfortable spaces to be explored. This is the making of diffractive data entanglements that push beyond normalised boundaries and practices in an adventure of uncharted exploration and a childhoodnature inquiry of what is yet to be found and known (Koro-Ljungberg et al., 2020).

The following eight principles of diffractive methodology arose from a deep exploration into some of the critical and fundamental works in diffractive methodologies to date (Barad, 2014; Hultman & Lenz Taguchi, 2010; Jackson & Mazzei, 2011; Lenz Taguchi, 2012; Mazzei, 2014; Murris & Bozalek, 2019; Palmer, 2011; Taylor, 2013). Through this exploration, I sought to find principles that would provide a foundation for my research to give it greater meaning through the research undertaken before it. The eight principles are outlined in Figure 5.1.

Data is lively

This principle explains the phenomena of the agency of data; that data is a material-discursive practice and an agentic force. Data is dynamic and exists in a state of flux. It is in constant intra-action with other data, other bodies, other matter. As such, data is relational and is a practice of relationality. This principle provides a shift from the thinking that data is "inert, passive 'stuff' we (humans) go out and 'collect', return with and then pore over to analyse, code and thematise" (Taylor, 2013, p. 691). Data's aliveness is an integral

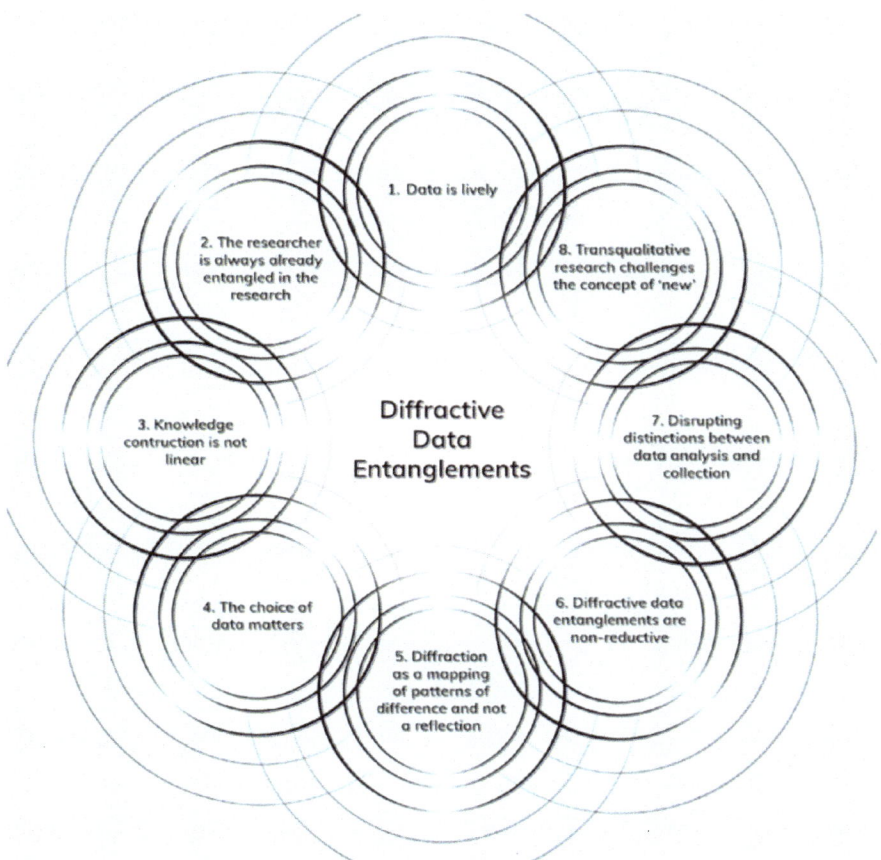

FIGURE 5.1   A visual presentation of the eight principles of diffractive data entanglements (adapted from Barad, 2014; Hultman & Lenz Taguchi, 2010; Jackson & Mazzei, 2011; Lenz Taguchi, 2012; Mazzei, 2014; Murris & Bozalek, 2019; Palmer, 2011; Taylor, 2013). *Illustration by Kelly @ Kelly Designs for the author.*

aspect of the research. Indeed, the researcher's role with and as the data is to listen to the data and give it voice. In the aliveness of data, independent of spacetime, data is returned to, reworked and reconceptualised (Barad, 2014). Through the process of *returning, reworking,* and *reconceptualising* the lesson participations, video-stimulated recall teacher conversations and visual-journal entries, data from the teacher participant engaged in a form of discourse with each other. This discourse of the data or, dance of the data, demonstrates the relationality between places, spaces and times. Diffraction is an iterative process that grants the data its inherent agency (its materiality) within the scaffolds of the ethnographic methods that it is entangled with. This means that the data is material and therefore, is discursive (it speaks) and when data is known, it constitutes a rich way of doing data that focuses on the human/nonhuman.

The researcher is always already entangled in the research: fracturing the subject-object binary

Research has been founded on many traditions that are based on quantitative thinking. This includes the perception that as a researcher, you can remain an objective "observer" of the data without influencing, impacting, shaping or shifting the data at all. Qualitative research has offered a different perspective in identifying that the researcher is an entangled part of the research; as, from a quantum level, they are never apart from it but are a part of it (Barad, 2007; Taylor, 2013). In addition, transqualitative research does not pretend the participants' data is pure and untouched; it highlights that interview data (for example) is already an interpretation as the participant has already "filtered, processed and interpreted each happening themselves – they are not the victim of some 'thing' happening to them" (Mazzei & Jackson, 2012, p. 3).

The entanglement of the researcher and the data in this study was particularly apparent through the diffractive analysis. As Barad (2007) proposes, diffraction is "a way of understanding the world from within and as part of it" (p. 88). In this book, my position as a researcher as an active agent in the research creation process does not attempt to hide, mask or curtail my subjectivity and indeed, highlights the value and strength of being deeply entangled with/as the data. Moreover, I have not sought to validate my data in an attempt to quell assumptions proposed by quantitative research regimes that hungover into qualitative research practices. The data-making was presented following the diffractive protocol.

Returning to the quote that opens the Fourth Turn in this book, Barad (2014) explains the sometimes seemingly elusive process of diffraction in research practices. Barad uses the earthworm metaphor literally "earthing/grounding" this process for (un)doing data and detailing the way earthworms intra-act with the earth to illustrate the way data can be played with too. The worm does not merely look at the soil as a passive observer; rather it is *a part* of the soil, altering its form and composition. This can be similarly said for the researcher (as the worm) and the data (as the soil). The worm (researcher) does not reflect on the soil as a passive object of the past but iteratively turns and re-turns the soil (data) over and over again, a "multiplicity of processes" of becoming engaged through diffractive patterns. The data is active and dynamic; it is reconfigured with each turn. The researcher plays a part in the performance of the data that cannot be denied. The researcher is always already entangled with/in/through the data in a multitude of ways that are integral to the diffractive data entanglements. In adopting this thinking, the ensuing practice acknowledges the impact the researcher always already has on the data. There is no replication or repetition of research data to reduce outliers and confirm and conform results. The researcher holds response-ability in enacting the agential-cuts to the data and as such, at any point in spacetime that the research is "repeated", the data would be different. The diffractive data entanglements are an ethico-onto-epistemological practice that is "about accountability to marks on bodies, and responsibility to the entanglements of which we are a part" (Barad, 2012, p. 52).

An ethico-onto-epistemology works to fracture divisive binary practices in research to consider ways that demonstrate the entanglement of subject and object; two research entities that have traditionally been considered separate. The entanglement of the researcher in the research data, in a move away from the subject-object binary, is important in understanding how diffractive methodological approaches differ from interpretivist analysis processes: diffraction is itself *an entanglement of mapping data movements in spacetime*. Diffraction seeks to mark the patterns, the ripples of difference, that the data makes. Rather than looking back over data to identify the areas of sameness, diffraction marks the patterns created when the data intra-acts with an interference: the diffraction gratings.

In addition, diffraction can make "agential cuts" – opening up the possibilities that "bring inventive provocations; they are good to think with. They are respectful, detailed, ethical engagements" (Barad, 2012, p. 50). The process of diffraction is "suggestive, creative and visionary" rather than reflective practices that seek to identify sameness (Barad, 2012, p. 50). This inquiry adopted three diffraction gratings with which to view the data (see Figure 4.1 in the Fourth Turn). The way data intra-acted with the concepts proposed (the diffraction gratings) and then further intra-acted with each other, through the diffraction ripples, aptly illustrated a metaphor for the data-making process that adopts the notion of *mapping data movements in spacetime*.

Knowledge construction is not linear

When thinking with diffraction, it can be useful to consider a tangible example such as the visual of diffractive movements as they occur in nature. Most commonly, the diffractive ripples that occur in and through water come to mind. Consider the photos showing the diffraction of ocean waves presented by Barad (2007, p. 75). The ocean is a tumultuous and non-linear arrangement of water that gathers to flow through a diffraction grating such as an opening in a barrier represented in the images as a gap in a rock wall. The large volume of water pushing through the opening creates a diffractive patterning where water diffracts outwards in a multitude of directions and equally flows back when the waves retreat. The water and ripples – like data – intra-act and affect each other. This intra-activity is the seen and unseen flows of energy that are always already occurring between human and nonhuman matter – this includes the human participants as well as the nonhuman materiality. Data is considered independent of spacetime and through the relationality of materiality (Barad, 2007). Diffractive data entanglement-making is therefore based on relationality, not linearity, which is inherent in the intra-activity of matter.

The choice of data matters

Diffractive ethnography enacts a responsibility in the researcher. The researcher is responsible for making the agential cuts to determine *what counts* when it comes to data. Taylor (2013, pp. 691–692) expands on the "agential cut" (as described by Barad, 2007),

> An analytic practice which both separates out "something" – an object, practice, person – for analysis from the ongoing flow of spacetimemattering, but which, at the same time as separating and excluding, entangles us ontologically with/in and as the phenomena produced by the cut we make.

As data is agentic and lively (see the first principle), "data have their ways of making themselves intelligible to us", where data "makes you feel kind of peculiar" and "agency feels distributed and undecidable, as if we have chosen something that has chosen us" (MacLure, 2013, p. 661). MacLure (2013) describes this as the data that "glows" and "glimmers": the researcher's embodied and affective response to the data. The researcher is then response-able for listening and tuning into the data to enact a bodily, affective response. Such responsibilities enable responsivity with and to the data – an intra-active dance of data with/ as the researcher.

Diffraction as a mapping of patterns of difference, and not a reflection

Reflective and reflexive methodologies attempt to separate the object at a distance from the world of which they are inevitably a part. A reflection describes a replication of the same, a mirror "image or representation". In traditional quantitative and science-based research methodologies, knowledge is "understood as a reflection of culture, rather than nature" (Barad, 2007, p. 86). Ethnography generally adopts reflexivity as a critical practice to ensure data validity (Hammersley & Atkinson, 2007). However, diffractive methodologies trouble reflexivity by only accounting for the "arrangement between objects, representations, and knowers" (Barad, 2007, p. 86). In addition, "reflexivity, like reflection, still holds the world at a distance" (Barad, 2007, p. 87), from an assumed position that data has been correctly interpreted by the researcher to produce an accurate reflection of reality. Haraway (1997, as cited in Barad, 2007) disrupts the concept of reflexivity,

Reflexivity has been recommended as a critical practice, but my suspicion is that reflexivity, like reflection, only displaces the same elsewhere.... What we need is to make a difference.... Diffraction is an optical metaphor for the effort to make a difference in the world.... Diffraction patterns record the history of interaction, interference, reinforcement, difference. Diffraction is about heterogeneous history, not about originals. Unlike reflections, diffractions do not displace the same elsewhere, in more or less distorted form ... rather, diffraction can be a metaphor for another kind of critical consciousness ... one committed to making a difference and not to repeating the Sacred Image of Same.

Since Haraway, and then Barad, founded diffraction as a methodological approach, it has seen a slow but steady rise in educational research (see, for example, Blom & Crinall, 2020; Bozalek & Zembylas, 2017; Crinall, 2017; Davies et al., 2013; Lather, 2013; Lenz Taguchi, 2012; Mazzei & Jackson, 2012; Mitchell, 2017; Murris & Bozalek, 2019; van der Tuin, 2011; Verlie, 2018) as researchers are aware of the limitations imposed when research is reduced and condensed into processes of reflection and refraction that are not applicable in all contexts. As such, diffraction seeks to make differences visible, to make differences known, and to make differences matter.

Diffractive data entanglements are non-reductive

Through exploring the selected extant research on diffractive methodologies, it was brought to attention that diffraction enables the expansion of data, where data-making practices are generative. This is the very nature of diffraction. Diffraction does not seek to condense, reduce, categorise or theme data in contrast to many data analysis processes that "crunch data" through a process where the "raw data are examined closely, passed, coded, counted, tallied, and summarized" (LeCompte & Schensul, 2012, p. 14). This commonly used research practice is designed to reduce and summarise data to bring order to the data, through coding and finding themes.

Diffractive data entanglements, then, take responsibility for expanding data through generative, data-making practices where "each moment is an infinite multiplicity" (Barad, 2014, p. 169). Diffractive data entanglements intra-act with the data to explore what is possible, bring attention to the quietened or silenced voices and intra-act with the reader to ignite further data-making practices.

Disrupting distinctions between data collection and analysis

Through the ethico-onto-epistomology binary-fracturing processes, it became necessary to disrupt isolated processes of data collection and data analysis. From this troubling, diffractive data entanglements emerged as a different way to approach the always already intra-active processes involved in data-making. This clearly defined practice of "diffractive data entanglements" in response to a diffractive ethnography provided the necessary setting for transqualitative research – as an overarching methodology – to emerge. Through attempting to define data as being collected or analysed, I realised the inherent conflict given my ethico-onto-epistemology positioning. Data was everywhere and everything. Data was interpreted by the teacher participant before being interpreted by the researcher participant. Not formal interpretations, but acknowledging the intra-active and iterative processes always already taking place. Diffractive data entanglements describe how data collection and analysis processes overlap and become one in a messy, iterative and dynamic process of material data-making. Data, like other phenomena, are not "independent objects with inherent boundaries and properties" (Barad, 2007, p. 25), but an ongoing intra-activity.

The data entanglements are moments in spacetime where a single intra-action is a multiplicity of parts contributing to the data's becoming. In this study, this included, but was not limited to, for example, my intra-actions with the lesson participations, video-stimulated recall conversations, photos, video recordings, the visual-journal entries along with the material-discursive practices that inform my movements such as being a PhD candidate, a teacher, a mother, a woman, my age, my socio-economic position, my relationship with nature.... All of these "parts" – the material-discursive forces – also inform other bodies in the study and that may be considered to *identify* me as an individual "self", influence movements and ethico-onto-epistemologies. Data entanglements are made under the impression of these influences.

Transqualitative research challenges the concept of "new": Time is diffracted as past, present, future collapse through the concept of spacetimemattering

As has been elucidated so far, the researcher holds response-ability for the agential cuts which are the "boundary-making practices" (Barad, 2007, p. 136) that become the data. The researcher is responsible for deciding *what* data is included and *when* the data-making practice ends. In enacting such mark-making practices in a diffractive ethnography as transqualitative research, there is no "new"; there is only a "(re)configuring of patterns of differentiating-entangling. As such, there is no moving beyond, no leaving the 'old' behind.... **There is nothing that is new; there is nothing that is not**

new" (Barad, 2014, p. 168; emphasis in original). This quote from Barad (2007) is indeed an example in and of itself. Thinking philosophically with King Solomon who claimed, "What has been will be again, what has been done will be done again; there is nothing new under the sun" (Ecclesiastes 1:9). Barad (2014) presenting the modern quantum science behind the philosophy presented by King Solomon over 1,000 years prior; again, in itself a spacetime fracture.

The concept of "new" presented here, challenges the way language has used the word "new" in everyday vernacular. For example, the "new waves are formed" may be used to describe the process of diffraction waves being generated and formed. Considered another way, classical physics describes the notion of "nothing being new" through *The Law of Conservation of Energy*: where matter is neither created nor destroyed but transformed or transferred from one form into another (Cleveland & Morris, 2015). Barad's (2014) example of the earthworm returning to the soil is a great example of this. In addition to this, quantum mechanics explains the de-linearisation of time, "there is no absolute boundary between here-now and there-then" (Barad, 2014, p. 168). When time is diffracted, anything is possible.

Diffractive data entanglements are dynamic and as such, always shifting, changing, and evolving. The data is presented as a moment in spacetime that explores ways of resisting the imposts of time and space and provides a context for the researcher and participant to revel in the revelations that can be seen with each (re)turn to the data.

Conclusion

Diffraction was described as a way of mapping where *the effects of difference occur* through diffractive patterning and the entanglement of matter through a relational ontology and as a methodological approach through diffractive ethnography. Diffractive ethnography privileges the materiality and relationality of all matter, not just human bodies. Diffractive ethnography takes conventional ethnography that studies people, society, and culture and applies a posthuman, diffractive turn to study phenomena. As such, diffractive ethnography questions the nature of entanglements and phenomena through the material, relational ontology. Diffraction, as a phenomenon, considers the materiality and relationality of matter that decentres the human and challenges binary thinking. Therefore, diffractive ethnography looks to the flows and patterns of data rather than cause and effect. Data presentation is also troubled as diffractive ethnography resists representation. Diffractive ethnography explores the entanglement of the human nonhuman including nonhuman nature and material discursive forces. In enacting a diffractive ethnography, the research in this book contributes to the ever-growing body of research utilising this methodological approach, while being a forerunner in applying this in the environmental education context.

The eight principles of diffractive data entanglements have emerged from a review of the extant literature that has previously used diffraction as a methodological approach. The eight principles were identified as:

1. Data is lively.
2. The researcher is always already entangled in the research.
3. Knowledge construction is not linear.
4. The choice of data matters.
5. Diffraction as a mapping of patterns of difference and not a reflection.
6. Diffractive data entanglements are non-reductive.

7. Disrupting distinctions between data analysis and collection.
8. Transqualitative research challenges the concept of "new".

Diffractive data entanglements describe the way data makes itself known through the entangled practice of data collection and data analysis. Data is always already entangled as the processes of data collection and data analysis are inherently in constant intra-action. Moreover, I argued that data are not passive or inert but lively, affectual, and relational, where relationality lies in the responsibility of the entanglement.

Note

1 I acknowledge that often in ethnography, the people being studied are called "informants". However, the term "participants" has been adopted in this study to denote the role and position of the human actants more accurately in this study. This concept is explored more fulsomely in the Sixth Turn.

References

Barad, K. (2007). *Meeting the universe halfway: Quantum physics and the entanglement of matter and meaning*. Duke University Press.

Barad, K. (2011). Erasers and erasures: Pinch's unfortunate "uncertainty principle". *Social Studies of Science, 41*(3), 443–454.

Barad, K. (2012). "Matter feels, converses, suffers, desires, yearns and remembers": Interview with Karen Barad. In R. Dolphijn & I. van der Tuin (Eds.), *New materialism: Interviews & cartographies* (pp. 48–70). Michigan Publishing.

Barad, K. (2014). Diffracting diffraction: Cutting together-apart. *Parallax, 20*(3), 168–187. https://doi.org/10.1080/13534645.2014.92 7623

Blom, S. M., & Crinall, S. (2020). Growing communities in a garden undone: Worldly justice, withinness and women. *Genealogy, 4*(2), 42.

Bozalek, V. (2017). Slow scholarship in writing retreats: A diffractive methodology for response-able pedagogies. *South African Journal of Higher Education, 31*(2), 40–57.

Bozalek, V., & Zembylas, M. (2017). Diffraction or reflection? Sketching the contours of two methodologies in educational research. *International Journal of Qualitative Studies in Education, 30*(2), 111–127.

Cleveland, C. J., & Morris, C. (2015). Conservation of energy. In C. J. Cleveland & C. Morris (Eds.), *Dictionary of energy* (2nd ed.). Elsevier.

Crinall, S. (2017). *Blogging art and sustenance: Artful everyday life (making) with water.* Western Sydney University (Australia).

Davies, B. (2021). *Entanglement in the world's becoming and the doing of new materialist inquiry.* Taylor & Francis Group.

Davies, B., De Schauwer, E., Claes, L., De Munck, K., Van De Putte, I., & Verstichele, M. (2013). Recognition and difference: A collective biography. *International Journal of Qualitative Studies in Education, 26*(6), 680–691.

Denzin, N. K., & Lincoln, Y. S. (2011). *The Sage handbook of qualitative research.* Sage.

Elfström Pettersson, K. (2017). Teachers' actions and children's interests: Quality becomings in preschool documentation. *Nordisk Barnehageforskning, 14*(2), 1–17. https://doi.org/10.7577/nbf.1756

Ellis, C., Adams, T. E., & Bochner, A. P. (2011). Autoethnography: An overview. *Forum: Qualitative Social Research, 12*(1), Article 10. https://www.qualitative-research.net/index.php/fqs/article/view/1589/3095

Gobo, G., & Marciniak, L. T. (2016). What is ethnography? In D. Silverman (Ed.), *Qualitative research* (pp. 103– 120). Sage Publications.

Gullion, J. S. (2018). *Diffractive ethnography.* Routledge. https://doi.org/https://doi.org/10.4324/9781351044998

Hammersley, M. (2018). What is ethnography? Can it survive? Should it? *Ethnography and Education, 13*(1), 1–17. https://doi.org/10.1080/17457823.2017.1298458

Hammersley, M., & Atkinson, P. (2007). *Ethnography: Principles in practice.* Routledge.

Haraway, D. (1992). The promises of monsters: A regenerative politics for inappropriate/d others. *Cultural studies* (pp. 295–337). Edinburgh University Press.

Holman Jones, S., Adams, T., & Ellis, C. (2013). *Handbook of autoethnography.* Left Coast Press.

Hultman, K., & Lenz Taguchi, H. (2010). Challenging anthropocentric analysis of visual data: A relational materialist methodological approach to educational research. *International Journal of Qualitative Studies in Education, 23*(5), 525–542.

Ingold, T. (2014). That's enough about ethnography! *Hau: Journal of Ethnographic Theory, 4*(1), 383–395.

Jackson, A. Y., & Mazzei, L. (2011). *Thinking with theory in qualitative research: Viewing data across multiple perspectives.* Routledge.

Kaiser, B. M., & Thiele, K. (2014). Diffraction: Onto-epistemology, quantum physics and the critical humanities. *Parallax, 20*(3), 165–167. https://doi.org/10.1080/13534645.2014.927621

Koro-Ljungberg, M., Tesar, M., Hargraves, V., Sandoval, J., & Wells, T. (2020). Porous, fluid, and brut methodologies in (post)-qualitative childhoodnature inquiry. *Research handbook on childhoodnature: Assemblages of childhood and nature research* (pp. 277–294). Springer International Publishing.

Larson, M. L., & Phillips, D. K. (2013). *Searching for methodology: Feminist relational materialism and the teacher-student writing conference.* Faculty Publications.

Lather, P. (2013). Methodology-21: What do we do in the afterward? *International Journal of Qualitative Studies in Education, 26*(6), 634–645.

LeCompte, M. D., & Schensul, J. J. (2012). *Analysis and interpretation of ethnographic data: A mixed methods approach.* Rowman Altamira.

Lennon, S. (2017). Re-turning feelings that matter using reflexivity and diffraction to think with and through a moment of rupture in activist work. *International Journal of Qualitative Studies in Education, 30*(6), 534–545. https://doi.org/10.1080/09518398.2016.1263885

Lenz Taguchi, H. (2012). A diffractive and Deleuzian approach to analysing interview data. *Feminist Theory, 13*(3), 265–281. https://doi.org/10.1177/1464700112456001

Lloro-Bidart, T. (2018). A feminist posthumanist multispecies ethnography for educational studies. *Educational Studies, 54*(3), 253–270.

MacLure, M. (2013). Researching without representation? Language and materiality in post-qualitative methodology. *International Journal of Qualitative Studies in Education, 26*(6), 658–667. https://doi.org/10.1080/09518398.2013.788755

Malone, K. (2020). Re-turning childhoodnature: A diffractive account of the past tracings of childhoodnature as a series of theoretical turns. *Research handbook on childhoodnature: Assemblages of childhood and nature research* (pp. 1–31). Springer International Publishing.

Mazzei, L. A. (2014). Beyond an easy sense: A diffractive analysis. *Qualitative Inquiry, 20*(6), 742–746.

Mazzei, L. A., & Jackson, A. Y. (2012). Complicating voice in a refusal to "let participants speak for themselves". *Qualitative Inquiry, 18*(9), 745–751.

Mellander, E., & Wiszmeg, A. (2016). Interfering with others-re-configuring ethnography as a diffractive practice. *Kulturstudier, 7*(1), 93–115.

Mitchell, V. A. (2017). Diffracting reflection: A move beyond reflective practice, *Education as Change, 21*, 165–186. http://www.scielo.org.za/scielo.php?script=sci_arttext&pid=S1947-94172017000200010&nrm=iso

Moxnes, A., & Osgood, J. (2019). Storying diffractive pedagogy: Reconfiguring groupwork in early childhood teacher education. *Reconceptualizing Educational Research Methodology, 10*(1), 1–13.

Murris, K. (2017). Reading two rhizomatic pedagogies diffractively through one another: A Reggio inspired philosophy with children for the postdevelopmental child. *Pedagogy, Culture & Society, 25*(4), 531–550.

Murris, K. (2020). Posthuman child and the diffractive teacher: Decolonizing the nature/culture binary. In A. Cutter-Mackenzie, K. Malone, & E. Barratt Hacking (Eds.), *Research handbook on childhoodnature: Assemblages of childhood and nature research* (pp. 1–25). Springer International Publishing. https://doi.org/10.1007/978-3-319-51949-4_7-2

Murris, K., & Bozalek, V. (2019). Diffracting diffractive readings of texts as methodology: Some propositions. *Educational Philosophy and Theory, 51*(14), 1504–1517.

Murris, K., & Haynes, J. (2018). *Literacies, literature and learning: Reading classrooms differently.* Routledge.

Pacini-Ketchabaw, V., Taylor, A., & Blaise, M. (2016). Decentring the human in multispecies ethnographies. In C. A. Taylor, & C. Hughes (Eds.), *Posthuman research practices in education* (pp. 149–167). Palgrave Macmillan.

Palmer, A. (2011). "How many sums can I do"? Performative strategies and diffractive thinking as methodological tools for rethinking mathematical subjectivity. *Reconceptualizing Educational Research Methodology, 2*(1). https://doi.org/10.7577/rerm.173

Pink, S. (2012). *Advances in visual methodology.* Sage.

Pink, S. (2015). *Doing sensory ethnography.* Sage.

Schneider, J. (2002). Reflexive/diffractive ethnography. *Cultural Studies? Critical Methodologies, 2*(4), 460–482.

Sehgal, M. (2014). Diffractive propositions: Reading Alfred North Whitehead with Donna Haraway and Karen Barad. *Parallax, 20*(3), 188–201. https://doi.org/10.1080/13534645.2014.927625

Taylor, C. A. (2013). Objects, bodies and space: Gender and embodied practices of mattering in the classroom. *Gender and Education, 25*(6), 688–703. https://doi.org/10.1080/09540253.2013.834864

Taylor, A., & Blaise, M. (2014). Queer worlding childhood. *Discourse: Studies in the Cultural Politics of Education, 35*(3), 377–392. https://doi.org/10.1080/01596306.2014.888842

Ulmer, J. B. (2017). Posthumanism as research methodology: Inquiry in the Anthropocene. *International Journal of Qualitative Studies in Education, 30*(9), 832–848.

van der Tuin, I. (2011). "A different starting point, a different metaphysics": Reading Bergson and Barad diffractively. *Hypatia, 26*(1), 22–42. https://doi.org/doi:10.1111/j.1527-2001.2010.01114.x

Verlie, B. (2018). Affective entanglements: learning to live-with climate change [PhD thesis, Monash University].

Visweswaran, K. (1994). *Fictions of feminist ethnography.* University of Minnesota Press.

6

THE SIXTH TURN

Data designing and methods making in posthuman educational research

Introduction

In this book, I have introduced a transqualitative inquiry (see the Fourth Turn) to explore how teachers perceive nature and how it informs their pedagogy. Thinking with the posthuman (see the Third Turn) provides opportunity to reconceptualise qualitative methods such as observations and interviews differently. In doing so, qualitative methods are reframed from observations into lesson participations and interviews into conversations, to align with posthuman thinking and therefore, a transqualitative methodology. As Murris (2020, p. 10) argues,

> Much depends on how the words and concepts we are so used to in research are de(con)structed (Barad, 2017) and reconfigured. After all, well-known terms, concepts and experiments can fall into new relationships with one another or, perhaps better put, in other relationships.

If researchers and methodologists are to embrace the transqualitative realm that seeks to create "new" (emphasis added here to

DOI: 10.4324/9781032703473-6

signify that this is a troubled notion that has been discussed in previous Turns) and experimental ways of doing data, this must also
include playing with and exploring how qualitative methods can
be reconceptualised in different and relational ways.

A transqualitative methodology aims to support diffractive
data entanglements – the entangled process of data collection
and analysis – that are deeply embedded in the posthuman.
Transqualitative inquiry embraces the tension of working with
human-centred methods posthumanly. Disengaging from the
tendency to lean into practised and well-rehearsed qualitative
methodologies is a difficult undertaking. Non-traditional methodologies, such as transqualitative, are designed to disrupt and disturb business-as-usual research approaches to open up different
ways of thinking and doing research (MacLure, 2013; Murris, 2020;
St. Pierre, 2011).

The design of data becoming

The research design underpinning the data presented in this book
is described through four discrete but intraconnected phases (see
Figure 6.1). Phases one, two and three of the research design
include diffractive ethnographic methods of (i) lesson participations, (ii) video-stimulated recall conversations with the teacher,
and (iii) visual-journaling. The fourth phase is the diffractive data
entanglements – the entanglement of data collection and analysis
processes.

The nature of these phases of inquiry is overtly and purposefully human-centred to contextualise the research in the current
framing of teacher practices and pedagogies. The humanistic focus
of education systems is a fundamental tension of this work; that
is, research that is grounded in posthuman theories yet applied to

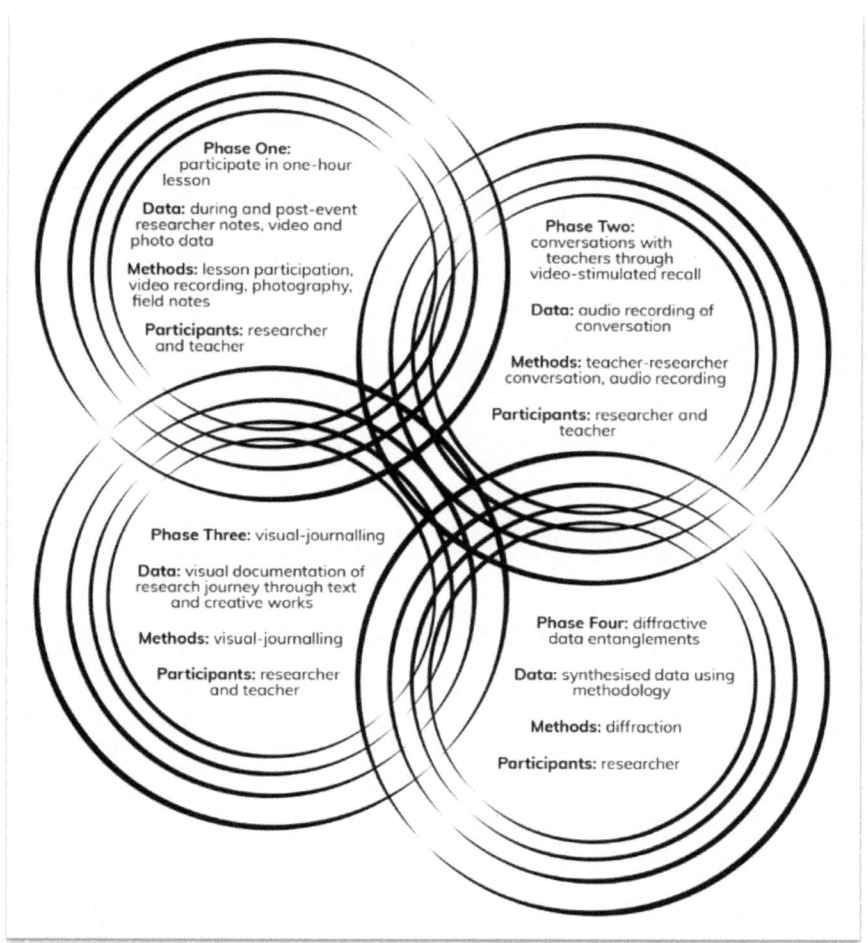

FIGURE 6.1   The four phases and entangled methods of this study. *Illustration by Kelly @ Kelly Designs for the author.*

human-centred methods. To trouble this tension, I brought voice and attention to the relationality of the material that is in constant discourse as part of the data becoming. The methods applied here are ethnographic, however, as was argued in the Fifth Turn, recent works in alternative genres of ethnography have enabled the emergence of different ways of "doing data" that account for a greater material focus in ethnographic approaches (Pink, 2015; Vladimirova

& Rautio, 2020). It is through the adoption of a diffractive ethnography, which has a core tenet of troubling binaries and emphasising the boundary-making practices of agential cuts (Barad, 2007), that the boundaries of qualitative research are pushed into a transqualitative methodology. This book adopts a diffractive ethnography to explore the potential of transqualitative research while also disrupting the traditional associations of ethnographic data with social culture to include nature, therefore troubling the nature-culture divide.

Phase four of the research involved pushing the data through theory – the diffraction gratings – in the continued process of making diffractive data entanglements. Diffractive data entanglements utilise the emergent data analytical tool of diffraction to make data entangled with and through the Returning Learning theoretical framework (proposed in the Third Turn) which includes the concepts of material-discursive forces, affective atmospheres and childhoodnature. While the concepts of childhoodnature (see Blom, 2020; Cutter-Mackenzie-Knowles et al., 2020) and material-discursive forces (Barad, 2007; Brown et al., 2020) are gaining popularity in educational research, affective atmospheres is still largely unknown despite some pivotal contributions which have grounded the theory in this context (Murris & Haynes, 2018; Verlie, 2018; Verlie & Blom, 2021). In keeping with the posthumanist theoretical framing, making data with and through the Returning Learning theoretical framework pays attention to the materiality and relationality of matter, where matter includes everything, including the social and cultural practices (Barad, 2007) that may impact and influence perceptions of the everyday lives of individuals and nonhuman others (Koro-Ljungberg, 2015).

Although the making data process is described linearly here (see Figure 6.1 and Table 6.1), this research acknowledges and

TABLE 6.1 The methods and other considerations of the study

Phase	Focus	Participants and Description	Methodological Approach	Methods	Data Documentation	Data Entanglements (Phase 4)	Ethical Considerations	Timeline
1	To inform an understanding of the pedagogies that are applied by the teacher	Teacher Researcher Learning environment	Diffractive ethnography	Lesson participation Researcher notes Video recording Photography	Video recording Photographs Field notes (during and after)	Diffraction through the Returning Learning theoretical framework	Consent from classroom teacher; acceptance of invitation from Principal; opt-out option for parents Anytime opt-out	February 2020 (start term 1)–August 2020 (mid-term 3)
2	To inform an understanding of the teacher's choice of pedagogy and perception of nature	Teacher Researcher	Diffractive ethnography	Conversation by video-stimulated recall Follow-up conversations	Audio recording	Diffraction through the Returning Learning theoretical framework	Consent from classroom teacher; acceptance of invitation from Principal Anytime opt-out	February 2020 (start term 1)–August 2020 (mid-term 3)
3	To create greater data depth and a method of triangulation	Teacher Researcher Journal and other materials	Diffractive ethnography	Visual-journaling	Visual works	Diffraction through the Returning Learning theoretical framework	Consent from classroom teacher; acceptance of invitation from Principal. Anytime opt-out	February 2020 (start term 1)–August 2020 (mid-term 3)

problematises the represented words, models and theories for the limitations they place on the reality of the entangled, non-linear way that research and knowledge is generated (Gullion, 2018; Koro-Ljungberg, 2015).

The four phases of the diffractive data entanglements are summarised here:

- Phase one was the lesson participations. The lesson participations were generally one hour in duration and involved the recording of video, photography, and field notes. The focus of the lesson participation was the teacher's pedagogical practices and the materiality of the classroom or learning space to explore the intra-actions.
- Phase two involved teacher-researcher video-stimulated recall conversations. This data was recorded through audio-recording and later, transcription of the data. The conversation was designed to understand the teacher's perception of the lesson and provided a space for discussing their decision-making processes.
- Phase three of the data entanglements was visual-journaling. This practice was undertaken by the researcher and the teacher participant was invited to create their own visual-journal. This practice provided another method to explore perceptions and views of nature, pedagogies, and local and or global happenings that attend to and influence the diffractive data entanglements.
- Phase four was making the diffractive data entanglements through the diffraction grating using the concepts of childhoodnature, affective atmospheres and material-discursive practices.

This process of four phases is presented as a flow-chart of diffractions (Figure 6.1) and in table format (Table 6.1).

Data-making methods

Diffractive ethnography (see the Fifth Turn) utilises the theories offered by diffraction entangled with the conventions of ethnography. I now explore the methods used as part of a diffractive ethnographic approach.

Diffractive ethnographic methods

The diffractive ethnographic methods – also known as the data-making techniques – applied in this research project, were adapted from the data collection traditions of ethnography, seeking to understand the views and visions of the world from the teacher participant (Denzin & Lincoln, 2011). However, this book also sought to understand the materiality and relationality of matter that intrinsically forms part of the data. In response to this challenge, I considered how existing ethnographic methods would be conceptualised through a posthuman framework. As Gullion (2018) proposes, diffractive ethnography is well placed to still use the "tools of social science inquiry (with its interviews and observations)" (p. 160) reframed with posthuman thinking that opens up inquiry to different ways of intra-acting with data.

To apply this theory to practice, transqualitative inquiry, as an overarching methodology, enables a dwelling in the tension of qualitative methods with posthuman theories. Gullion (2018) states "none of the 'methods' … are prescriptive … many exciting new forms of research will emerge in this [diffractive ethnographic] transition" (p. 28). This notion was similarly argued by

Murris (2020) in reconsidering how "well-known terms, concepts and experiments" (p. 10) from qualitative research can be reconceptualised in different methodological approaches.

As presented in Figure 6.1 and Table 6.1, the methods used in this research were (i) lesson participations (including photography and video recording), (ii) video-stimulated recall conversations with the teacher (audio recording and transcription), and (iii) visual-journaling. These methods sought to transform ethnographic data collection approaches into the posthuman and in doing so, challenge the binaries and separatist notions of collection and analysis. Instead, I propose that these data-making processes are already entangled with theory in their becoming.

Data-making processes problematise the human-centred focus of ethnography and bring attention to the materiality and relationality of matter. That is, highlighting the importance of all materiality and not *just* human bodies and the entangled nature of data-making practices. In a diffractive ethnography, attention is not only given to understanding humans through their behaviours and cultures. The human is decentred, and the human nonhuman is centred. That is, the human and nonhuman are equally considered and understood through and with their materiality, intra-actions and processes (Gullion, 2018; Larson & Phillips, 2013; Lenz Taguchi, 2010). The movement of the human body as material is given attention. Thus, with each data-making event (e.g., the lesson participation, video-stimulated recall conversation, and visual-journal entry), the researcher is ignited to become a genuine part of the data's agency and becoming; that is, the researcher is not an uninfluential observer but is actively taking part and acknowledging their role in the data entanglement.

Lesson participations including audio and video recording: rethinking ethnographic participant observations through the posthuman

In ethnography, participant observation is a commonly used method. Participant observation is a research setting where a participant is studied by a researcher (Atkinson & Hammersley, 1998). Participant observation provides a research method where the researcher can understand the lives and cultures of humans by closely observing the details of their lives (Lofland & Lofland, 2006; Musante & DeWalt, 2010). These tenets of participant observation are still apparent in diffractive ethnography, however, rather than observing the participant in a lesson from an outside position as an observer, the researcher is purposefully invited into the research situation. As such, diffractive ethnography challenges researchers to become different within themselves and be open to no longer being outside the research in a position where they can "experience the event differently" (Lenz Taguchi, 2010, p. 172); both physically and conceptually.

To achieve this difference, the researcher must become entangled in the research and not merely observe that data from the outside but from within. Therefore, in this study, humans were considered active participants in the continual becoming of the research; human participants and nonhuman participants were equally part of the data entanglements. Humans, as a kind of apparatus, "are not passive observing instruments. On the contrary, they are productive of (and part of) phenomena" (Barad, 2007). Ingold (2014) explains this phenomenon as "observing from the inside" and strongly argues that there can be "no observation without participation – that is, without an intimate coupling ... of observer and observed" (p. 387). In this book, I have adopted

Ingold's proposition literally to move from participant observation to lesson participation to glimpse all positions of the nature/culture of each lesson (Gullion, 2018; Ingold, 2014). In practice, I was positioned in the classroom in a location that did not seek to "hide" my presence, but one that worked in consultation with the classroom teacher to have the least impact on the flow and focus of the lesson and thus not interrupt or interfere with the students' learning. In some instances, this involved being introduced to the class, my research work explained, where the children engaged in asking me questions about what I was doing.

Participant observation involves making field notes to document the research data. Field notes are often used in combination with other ethnographic methods to enable a deeper and more comprehensive understanding of the culture being studied (Atkinson et al., 2001). During and following the one-hour lesson, I made some field and post-event notes of all the human and nonhuman participants and the movements and intra-actions between them in a field notes journal. It was found after the initial lesson participation, that immediate field notes were problematic while negotiating the technology, so post-event field notes were used for the remainder of the study. The lessons were photographed using an iPod to demonstrate the arrangement of bodies in the room. The lessons were also video recorded using an iPhone and a GoPro[1] primarily for use in the video-stimulated recall conversations but also to document further data for exploration and opportunities for data-making (Kellehear, 1993).

Video-stimulated recall conversations: reconceptualising ethnographic interviews through posthumanism

Ethnographic interviews are considered to be structured or semi-structured conversations or question-and-answer situations between the researcher and interviewee. Classically, ethnographic interview processes involve (i) thematising, (ii) designing, (iii) interviewing, (iv) transcribing, (v) analysing, (vi) verifying, and (vii) reporting (Atkinson et al., 2001). In this book greater focus is applied to the spontaneity of conversation that draws on the data from the lesson participation video and in light of the research questions, in a move away from thematising, analysing, and verifying. The term "conversation" is purposefully used to breakdown the power relations involved with the word "interview" enabling researcher and participant to work from more equal positions. "Conversation" was previously used by Lenz Taguchi (2012) in their study that analysed interview data diffractively to move away from formal interviews that seek to confirm sameness and from interpretive approaches that seek to understand the data.

Video-based methodological approaches are not new and have been used extensively in many fields of research for a variety of purposes (Cutter-Mackenzie et al., 2015; Harris, 2016). More recently, environmental education has pioneered the inclusion of children's voices through video-based methodologies (Cutter-Mackenzie et al., 2015), however, given the purpose of this study in exploring teachers' perceptions, the teacher's voice is the focus of the conversational data through video-stimulated recall. Video-stimulated recall has been used in many disciplines of educational research and by many researchers over the last two decades (see, for example, Cutter-Mackenzie et al., 2015; Day, 1998; Harris, 2016; Morgan, 2007; Muir, 2010; Pink, 2012a; Powell, 2005; Rowe, 2009). This method involves video-recording the lesson where the researcher is participant, followed by the researcher and teacher watching the video and conversing about any thoughts, feelings or ideas that arise. The reason for engaging in video-stimulated recall as a method for engaging in conversations with the teacher is that it has been found to (i) enhance reflections, promote collaboration between teacher and researcher, provide a context and focus for teachers to articulate their thinking and feelings (Muir, 2010; Powell, 2005); (ii) enable teachers to observe their lessons from a new perspective, provide increased opportunities for the teacher to guide the conversation and disclose relevant events in the video that were meaningful for them (Rowe, 2009); and (iii) provide a platform for authentic conversations about a particular subject (Cutter-Mackenzie et al., 2015). Although Murris and Haynes (2018) did not explicitly label their use of video in their study as video-stimulated recall, they did watch the recordings of lessons with teachers to further close any researcher/teacher power gaps by engaging the teacher in authentic researcher practices to enable a slowing down of what happens in real-time for a more detailed and deeper consideration.

As discussed by Cutter-Mackenzie et al. (2015), some of the key ethical considerations in adopting a video-based methodology, which also applies to this video-stimulated recall method, include "consent, researcher/participant relationships, confidentiality and protection, and data interpretation and analysis" (p. 2) (see further discussion under ethical considerations).

Caitlyn, who took part in the lesson participations, was invited to take part in two non-formal conversations through video-stimulated recall. This tool was used to elicit an understanding of Caitlyn's observations and perspectives of her lesson, her perception of nature, environmental education in general and/or anything else she wanted to discuss that she felt was relevant to the research. This approach to data differs from structured interviews, as the teacher and researcher informally discuss what the video provokes, and provides a scope for unintended and unplanned conversations (Moxnes & Osgood, 2019). Moreover, the use of "conversations" acknowledges the subjectivity of the data and that the researcher can never really adopt an objective position through which to interpret raw interview data. As described by Jackson and Mazzei (2011), the interviewee has "filtered, processed and already interpreted" themselves (p. 3); as such, the data is always already entangled. The interviewees are not the victims of some "thing" happening to them as they have already made meaning of their experience through what they choose to share and what they choose to silence (Jackson & Mazzei, 2011, p. 3).

The lesson participation video was mostly utilised to prompt conversation about the pedagogies that each teacher was enacting. The length and frequency of the engagement with the video recording varied heavily and was dependent on each participant. The conversations were audio-recorded with the teacher's consent so I could transcribe and re-listen to the recordings as, when and if needed.

Video

Since the turn of this century and the millennium, video has been heralded as the "new" visual method and noted for its increased use by researchers, including ethnographers (Aarsand & Forsberg, 2010). This rise in interest and use has also been experienced in educational research using video for the purpose of video-stimulated recall interviewing for its "insightful and useful data of examining the way people experience a specific event of interaction in education" (Nguyen et al., 2013, p. 1). In justifying the use of, contextualising and exploring ethical positions, video as a visual method is often used interchangeably with photography (Pink, 2012b).

In practice, two video cameras were set up for each video-recording aimed to capture the teacher's movements and practices as a relational performance in intra-action with other human and nonhuman actants. The GoPro camera was set up at the front of the class (Nguyen et al., 2013; O'Brien, 1993) and the iPhone camera was set up at the back of the room where I was positioned. These positions were at times difficult to maintain given the way that teachers move around in a classroom setting and therefore, the two cameras were positioned in a way to complement each other and to pick up the actions, movements and practices of the teacher participant across the whole classroom.

Photography

Photography, as one form of a visual method in ethnography, has burgeoned over the last two to three decades, becoming commonplace in ethnographic practice (Pink, 2020; Schwartz, 1989). Originally considered as a way to either represent data or to generate data (Schwartz, 1989), this aspect of photography is problematised by Pink (2020) who acknowledges the theoretical and methodological spaces that photography is being pushed into such as through the post-qualitative tenet of non-representation (Thrift, 2008). For example, Malone (2018) utilises photography as a visual method for young people to gather information about their lives. Although this study by Malone (2018) adopts a posthumanist/vital materialist theoretical stance, it also adopts ethnographic photography as a visual method used as photo-elicitation for generating verbal commentary (Schwartz, 1989, p. 151). Malone's (2018) study does, however, challenge the traditional ethnographic focus on human culture and society by including the nonhuman (e.g., dogs).

Ethnographers also use photography for reflexivity in an attempt to nullify the impacts of subjectivity, as photos are seen to yield an "unmediated and unbiased visual report" (Schwartz, 1989, p. 120). However, photography is also incredibly subjective yet subject to "multiple perceptions and interpretations", and herein lies the contradiction (Schwartz, 1989, p. 122). As transqualitative research blurs the subject-object binary and embraces the subjective as objective, photography was not used for this purpose in this study. As asserted by Pink (2012b), "subjectivity should be engaged with as a central aspect of ethnographic knowledge, interpretation and representation" (p. 126). The photographs in this study are considered part of the material-discursive practice, that is, through their materiality, they have a story to tell. As the researcher, I accept the responsibility of sharing this story and know the implicit influence of the *I* in doing so.

In ethnography, photography is oft used as the central method, however, in this research, photography was entangled with the other forms of data to enhance and explore the relationality between them (Pink, 2012b). Pink (2012b) highlights that even if visual research methods are focused on, they are not purely visual as "they cannot be used independently of other methods" (p. 124). In this book, photographs of the teacher in the lesson participations were taken before, during and after the lesson participations to document key moments, happenings or layouts in the classroom. Photographs were also drawn from the video recordings of the lesson participations. The photographs complement the lesson participation videos, field notes and video-stimulated recall conversation audios to add another layer and therefore greater depth to the data entanglements.

Field notes

Field notes are foundational in enabling ethnographers with a powerful way to document sensory experiences, "aha" moments and changes in understandings and perceptions as the data unfolds (Mills & Morton, 2013). Despite the myriad approaches of data documenting practices now available in light of the digital technological revolution such as "virtual, hyperlinked and diverse forms" (Mills & Morton, 2013, p. 78), I decided to use pen and paper for my field notes to maintain an uninterrupted and intimate interaction between me and the data – the human/nonhuman participants and happenings. Moreover, there was enough digital technology being utilised, through photography and video recording on two different cameras during the lesson participations and the audio recording during the video-stimulated recall conversations, that pen and paper was the most effective and unobtrusive way to mark key points and thinkings in the lesson participation and video-stimulated recall conversation process.

Defining and locating field notes is challenging due to the dynamic and varied way that researchers approach this method. It is accepted that the field notes-taking process is messy and informal – where notes and scribbles are taken during and or after the lesson participation event (Mills & Morton, 2013). Despite their undefined nature, field notes are treasured and highly regarded by ethnographers, even though not all researchers refer back to or even read their field notes again (Jackson, 1990). However, "note-taking allows and encourages individual reflexivity and creativity" while also prompting "later recall and elaboration" (Mills & Morton, 2013, pp. 80–81). Hammersley and Atkinson (2007) advocate for writing in a highly detailed form to limit the inference made by the researcher while Haraway (2003) argues for being situated and positioned *in* the data as "situated knowledges" that are about "communities, not about isolated individuals" (p. 122). This contention highlights the different views and positions between ethnographers about the way that field notes can and should be taken while also exposing more of the general subject-object binary in research.

In practice, I did not follow a strict or formulated approach to conducting field notes. Instead, I used a messy, impulsive and creative way to document the data. Wolfinger (2002) proposes that there are two modes of taking field notes, comprehensively as previously described by Hammersley and Atkinson (2007), or saliently, where the researcher documents the most noteworthy, interesting, and telling information. The latter mode described here most closely aligns with the method of field-note taking adopted in this study. Pedersen (2013) similarly asserts troubling traditional field notes by describing the "impossibility of zooethnographic representation" in a posthuman ethnographic study where she enacts practices of field notes in "vital and open-ended ways" (p. 728). Moreover, given the changes to research methods and advancements in technology over the last decade, and my integration of multiple methods simultaneously through the audio, photography, and video practices, I chose to only use field notes to mark key points that spoke and glowed (MacLure, 2013); the moments where data screamed to be heard, have a voice and be included in this research. I layered these notes by re-watching the videos and re-listening to the audios, adding to the data in a transqualitative re-turn to the data through turning the data over-and-over again.

Visual-Journaling as diffractive ethnographic method: reimagining ways of doing ethnographic data through adopting an arts-based educational research practice

Visual-journaling is not a common qualitative method; however, drawing has long been used by natural and social scientists along with ancient and First Nations cultures as a legitimate form of recording information. Since the 1990s, visual-journaling has been increasingly accepted as a research tool (Grauer & Naths, 1998; Messenger, 2016; Scott Shields, 2016). Over the last decade, there has been a notable increase in the validity of using visual-journaling in qualitative studies as part of the a/r/tography movement (La Jevic & Springgay, 2008). Visual-journaling is commonly utilised as a mechanism for (re)connecting pre-service teachers with arts-based theory and practice in an approachable, enjoyable and non-threatening way to support critical and relational meaning-making (La Jevic & Springgay, 2008; Sanders-Bustle, 2008).

Visual-journaling as applied in research supports reflexive practice and knowledge generation through a process of connecting senses, feelings, thoughts, and actions (Messenger, 2016). In particular, visual-journaling focuses on the sense of vision – as is common in arts practices and arguably, the world more broadly – as a sense-making mechanism to assist in meaning-making in the world (Eisner, 1992; Sanders-Bustle, 2008). Sanders-Bustle (2008) argued that this kind of meaning-making requires critical engagement in a subject and that this kind of critical engagement does not happen in isolation but relationally with creative engagement. The purpose of the visual-journal in the context of this study is to document the teacher's experiences beyond their immediate interactions with me in a way that allows their ideas and unfoldments to percolate and the expression of these to be recorded

(Sanders-Bustle, 2008); it is not a finished product of research, but evidence of the research process (Scott Shields, 2016).

Scott Shields (2016) relates the scientific process of diffraction to her experience of visual-journaling which has "the potential to create these same ripples of analytic and reflective thought" (p. 6) and is fittingly applied to this study. Moreover, this process of diffraction enables the participant to engage in a practice that demonstrates both process and product. The product provides visual context to allow the participant and audience to move through their developing understanding together while also demonstrating the process the participant went through in creating their product (Scott Shields, 2016). This process-product relationship of the visual-journal is suitably applied in transqualitative research for its creative and non-interpretive approach (Scott Shields, 2016).

At the commencement of this research, I began creating a visual-journal to support me with the processing of concepts and ideas and to provide a space for abstract thinking and knowledge generation. This practice continued throughout my study when I engaged in thinking about a topic or intra-action with the data and thus constituted part of the data entanglements. In addition to my visual-journal, each participant in the study was supplied with a visual-journal "pack" consisting of a visual-journal and some art-making materials. Similar to the purpose for myself as a researcher, the purpose for the teacher participant is to enable them to have a space to document their thinking and knowledge generation processes that are intentionally creative to support greater engagement with critical and creative thinking (La Jevic & Springgay, 2008; Sanders-Bustle, 2008). Caitlyn was supplied with visual-journals and art materials on her first lesson participation and the visual-journals were collected in a follow-up conversation later that year (term 4, 2020). This data is presented alongside the other data forms in the diffractive data entanglements in the Eighth and Ninth Turns.

Participants

In established anthropology studies – the birthplace of ethnography – human participants are often referred to as "informants" (Spradley, 2016). This term is problematic in perpetuating false hierarchies where there is someone – "the informant" – who has superior knowledge (Morse, 1991). Instead, the term "participants" is adopted in this study (Hammersley & Atkinson, 2007). "Participants" is commonly used in qualitative inquiries to denote an equal relationship between researcher and participant where the participant has greater control than an informant (Morse, 1991). In this book, I consider and sometimes refer to myself as "the researcher participant" as discerned from "the teacher participant" to demonstrate the intended equality of the participants.

The decision to have a small sample was to achieve depth of interrogation through intimate relationship building, meaningful conversations and to become entangled with the participant's school community (Creswell, 2018; Mills & Morton, 2013). While the original thesis contained four participants, to reduce the size of the thesis for this publication, it was necessary to select one participant's data to include. The importance of building trusting relationships with participants is a keystone of ethnographic research (Brewer, 2000; Dingwall, 1980; Parker, 2007). I took time to build relationships with the teacher participant and extended the relationship building by becoming entangled in the lands, communities and nonhuman worlds of each participant. As described by Mills and Morton (2013), there has been a focus in ethnography on "how best to 'be' around people rather than on relationships with them, and dwell little on the forms of empathy and involvement this might imply" (p. 68). Moreover, Mills and Morton advocate for time spent "hanging out" and highlight that "the value of getting entangled is best illustrated through examples of ethnographic work" (p. 68). The mechanism for "hanging out" was practised in numerous ways including school visits (pre-COVID), emails, newsletters and phone calls. As the participant was a previous student at the University where I teach, the relationship was already well-developed and had a strong basis of trust.

Purposeful participant selection is commonplace in qualitative research and suitably placed in ethnography (Reybold et al., 2013), compared to the random sampling counterpart of quantitative studies. Moreover, I argue that "purposeful selection" is also an appropriate method in transqualitative research as it is "a mechanism for making meaning, not just uncovering it" (Reybold et al., 2013, p. 700). The idea of making meaning is a form of knowledge generation that this study has adopted. The choice of year levels to limit to a teacher of either kindergarten, grade 1 and grade 2 was to conduct research into my area of interest and as a specific area of research that could be identified and that there is a discrete body of literature around.

This approach demonstrated the application of purposeful selection which has a (i) subjective focus that explores the tensions that exist when placing self in the inquiry "without manipulating interpretation and representation of data" (Reybold et al., 2013, p. 700), (ii) a quality procedure that considers the "sociocultural milieu in which the research is conducted" along with "information-rich cases that can give in-depth insight" (Reybold et al., 2013, p. 700), and (iii) an integrative method that views participant selection as a form of the entanglement of the data not just as a preparation prior to the research being conducted (Reybold et al., 2013). Through adopting a stance of participant selection as a method, it made sense to design my participant selection around my research questions which is particularly relevant to the area of focus of the early years of schooling. Essentially, this is the first agential cut into the data (Barad, 2007) that enacts a boundary-making practice to constitute what data is known as data.

The participant, Caitlyn (pseudonym) was responsible for choosing the topic and content of the lesson participation that I participated in. I had no prerequisite for the lesson topic, despite the focus of this study being in the field of environmental education. This enabled me to gain an authentic understanding of how environmental education is enacted in any of the key learning areas (disciplines). However, I note that a limitation of this study is that Caitlyn's decision-making process around lesson topic choice may have been influenced by the context of my research being focussed on environmental education.

In the posthuman turn, diffractive approaches enable the participants to be not merely the human actants but all the materialities of the space being researched (Barad, 2014; Lenz Taguchi, 2012; Taylor, 2013). On this, I acknowledge that not all are counted or numbered in this study. The presence of the nonhuman participants is recorded and documented as part of the convergence of ethnography (as a human-centred practice with participants) and diffraction (with a focus on materiality) (Gullion, 2018). The challenge in achieving this movement away from an anthropocentric focus, a focus which "puts humans above other matter in reality, that is, a kind of human supremacy or humanocentrism" (Hultman & Lenz Taguchi, 2010, p. 526), is acknowledged and the materiality of the space is picked up through the data entanglements in the final Turn through the data synthesis. I note that human perception is conditioned to view the human actants in a situation without equal focus on the nonhuman (Hultman & Lenz Taguchi, 2010).

Ethical considerations

Prior to making data, ethics approval needed to be obtained, from a Human Research Ethics Committee and the State Education Research Applications Process (SERAP) needed to be obtained as a requirement for conducting research in NSW government schools.

The primary concern of ethics is to ensure the safety, well-being, dignity, and privacy of human and nonhuman others being studied (Angrosino, 2006). The ethics procedures are guided by prior research as well as established ethical guidelines, such as the National Statement on Ethical Conduct in Human Research, 2007 (updated 2018, 2023) (Australian Government National Health and Medical Research Council [NHMRC] & Australian Research Council and Universities Australia, 2023).

Parents could opt their children out of being in the classroom at any time while the research project was taking place. The impact on the general classroom practices was designed to be minimal. The teacher participant was invited to direct any complaints concerning the research to the Human Research Ethics Committee, and details were provided in the information sheet.

Teacher's time is valuable and precious, and as such, one of the objectives of this research was to contribute to the teacher's personal and professional development. It was the researchers' heartfelt intention and firm commitment that the research process be worthwhile, meaningful, and reciprocally beneficial. This intent aligns with the National Statement on Ethical Conduct in Human

Research (Australian Government National Health and Medical Research Council [NHMRC], Australian Research Council and Universities Australia, 2023) which outlines that research must demonstrate respect, integrity, justice, and beneficence.

A non-reflexive, diffractive justification

The concept of reflexivity was born in an attempt to eliminate the "effects of the researcher on the data" by acknowledging the sociocultural positioning of the researcher (Hammersley & Atkinson, 2007, p. 15). Hammersley and Atkinson (2007) attest that rather than attempting to refute the researcher from the data, the process of understanding the researcher is what is most crucial. Through this process of highlighting the researcher's assumptions and effects on the research (Bozalek & Zembylas, 2017), the trustworthiness of the data is considered (Cutter-Mackenzie et al., 2015; Woolgar, 1988). It is an integral aspect of conventional ethnography, but noting it is an approach that is frequently and firmly rejected in posthuman theorising and methodologies, particularly by Barad (2007) and Haraway (1992).

These critiques are underpinned by the belief that (i) reflexivity practices perpetuate sameness by the nature of the science that reflection and refraction are founded, and (ii) reflexivity is dependent on fixed positioning, in other words, on representation (Bozalek & Zembylas, 2017). These ideas are problematic in diffractive theorising because approaches to doing data such as diffraction are concerned with the differences evident in research data, rather than the similarities or themes. Diffraction is interested in the differences that are co-constituted through research practices, not in more of the same. Diffraction argues for a rejection of linearity, of two points on the ground, of binaries. Diffraction disavows the notion of distance, which is the basis for reflexivity (Jackson & Mazzei, 2011), as in diffraction everything is already always; *here now, there then.*

Diffraction contends that the "entangled practices are productive, and who and what are excluded through these entangled practices matter: different intra-actions produce different phenomena" (Barad, 2007, p. 58). The "looking back" process of reflection in reflexivity could reduce opportunities for the data to make itself known. In doing so, "diffractive analysis goes beyond the idea of reflexivity and interpretation and produces new entangled ways of theorizing and performing research practices, co-constituting new possibilities of strengthening and challenging knowledges" (Bozalek & Zembylas, 2017, p. 14). This is significant as to push beyond the traditions of qualitative methodologies, it is important to discard the practices that no longer align with the theoretical positioning. Conventional reflexive practices do not fit the transqualitative inquiry framework opening up the raw possibilities and exciting emergences generated through the diffractive data process.

Data threads: the entanglement of be(com)ing (with) data

During the lesson participations (phase 1), I started with a GoPro camera at the front of the room, an iPhone camera from my position at the rear of the room and took photographs with an iPod. I also documented the lessons and conversations with field notes. However, after the initial lesson participations and conversations, I realised it was too challenging to navigate my position during classroom movements and happenings while observing the teacher and the two cameras in the case of the lesson participations, or maintaining my engagement and focus on the conversation with the teacher. Instead, I focused on post-event notes or audio recording my notes in my car after completing the lesson participation and video-stimulated recall conversation and transcribed them later. This refined practice aimed to record my experience in the lesson as close to the lesson happening as possible. This process of note-taking was significant in enacting an embodied processing and coalescing of the data, documenting a record of my initial thinking and informing the direction for the second round of participations and conversations (Mills & Morton, 2013).

In between the first and second rounds of lesson participations and teacher conversations, there were some large interruptions and uncertainties with the onset of COVID. During this period, I initiated a series of researcher-updates – akin to a two-page newsletter – which I sent to Caitlyn fortnightly that detailed my visual-journaling entries and theoretical thinking during the time. These two-page updates provided a touchstone for Caitlyn and enabled Caitlyn and I to get to know each other in different ways and from different per-spectives leading to greater depth of entanglement in the research process. When schools reopened in Term 3, 2020 the opportunity to reengage with the research presented itself.

On completion of the lesson participations and video-stimulated recall conversations, Caitlyn gave me her visual-journal. The final data set included:

- Lesson participations
  - Multiple video recordings (GoPro and iPhone), and
  - Photographs (iPod).
- Video-stimulated recall conversations
  - Audio-recordings, and
  - Transcriptions.

- Documentation including emails, newsletters, text messages, Facebook messenger messages, teacher's philosophies, newsletter updates.
- Teacher visual-journal entries, and
- Researcher visual-journal entries.

Diffractive data entanglements, enacted

To demonstrate the entanglement of data and researcher that a diffractive approach to data entanglements offers, Barad (2012) proposes a "suggestive, creative and visionary practice" for diffraction, "of reading diffractively for patterns of difference that make a difference" (pp. 49–50). The practices outlined below are not designed to be finite or conclusive, but like the earthworm preparing the soil for seeds to be sown and grown, as in transqualitative inquiry, the data entanglements (re)turn the data over and over again – within the limitations of language – to create "new" ground and "new" knowledge. In a physical sense, nothing is *new* here. The same stardust is cycled through a process, however, through my part as the researcher interrogating the research, the research is "new" – it has never been done before in this way, in this time, in this place: space-timemattering (Barad, 2017). Matter – the physicality, the light/dark energy and thinkings of the world and beyond – is cycled through, never created *new*, yet never the same as before.

Despite critiques of a "prescribed framework" for diffractive methodologies that claim it is "not desirable" (Murris & Bozalek, 2019), I outline the research process I have undertaken as a glimpse into how research processes can be useful in documenting how data and knowledge came to be known. This is not intended to be a prescribed framework but a mapping experiment to guide you, the reader, on my thinking and method of approaching data entangle-ments. As the researcher, my role in the data-making process is to demonstrate honesty and transparency; I am not attempting to hide or curtail my part in the research becoming. I am responsible for making the agential cuts – "an imaginative journey"– that disrupts tales of "scientific progress" (Barad, 2010, p. 244); this is key to dif-fractive data entanglements. I acknowledge that diffraction resists interpretation. Rather, it ignites and embraces generativity by pre-senting data that threads together – data that makes itself known.

The following four steps outline the enactment of the data entanglements:

1. On completion of phases 1 and 2 of the data-making process, I re-experienced the documented data through a deep dive into watching the video recordings of the lesson participations (data from phase 1), looking at the photos from the lesson partici-pations (data from phase 1), listening to the audio-recordings of video-stimulated recall conversations (data from phase 2), and entangling the teacher's and my visual-journal entries (data from phase 3). While re-engaging with the data I re-attuned to the concepts of childhoodnature, affective atmospheres, and material-discursive practices to explore how these concepts grow and expand the data in "new" and unexpected ways.

2. While re-engaging in the data, further field notes and visual-journal entries were documented to demonstrate how the data's agency emerges – how the data speaks and makes itself known and in doing so, how further data is generated.

3. The data with Caitlyn is mapped individually through pat-terning from the diffraction gratings (childhoodnature, affec-tive atmospheres, and material-discursive practices). This data

forms the Eighth and Ninth Turns; the diffractive data entanglements with Caitlyn.

4. On completion of this initial mapping of the data patterns through Caitlyn's data experiences and the diffraction gratings, the data is re-turned and synthesised through the extant environmental education literature, the Returning Learning theoretical framework and in response to the research questions in the Tenth Turn.

Conclusion

This Turn has introduced the design process of diffractive data entanglements through the four phases of this research study: (i) lesson participation, (ii) video-stimulated recall conversations with the teacher, (iii) visual-journaling, and (iv) making diffractive data entanglements. Data-making was enacted through the diffraction grating using the concepts of childhoodnature, affective atmospheres, and material-discursive practices. The linear four-phase process is troubled to emphasise the entangled, non-linear knowledge generation process.

The data-making, diffractive ethnographic methods entangled in these phases were then explored. The methods included lesson participations, video-stimulated recall conversations, video, photography, field notes, and visual-journaling. Lesson participations detail the movement away from lesson observations to consider the entanglement of the researcher and teacher participant in the materiality of the classroom space. As part of the lesson participations, the diffractive ethnographic methods of video, photography and field notes described how data were intra-acting as part of the data-making process. In addition, visual-journaling was a space for abstract thinking, conceptualising, grappling and generating that enacted knowledge generation through intentional creative practice. Through this exploration of the methods, I explained how they were practically implemented as part of my research design. In addition, I mapped out my process of entanglement of be(com)ing (with) data to describe the practical application of the methods. The study sought to ethically work with all human and nonhuman participants by ensuring research practices were meaningful and reciprocally beneficial.

The enactment of diffractive data entanglements was outlined in four steps, (1) deeply diving into the data, (2) further data-making through field notes and visual-journal entries, (3) data mapping, and (4) data synthesis. Which brought us to the completion of this Turn.

Note

1 A GoPro is a brand of camera that is popular in extreme sports for its capacity to capture video footage. For more information, see https://gopro.com/.

References

Aarsand, P., & Forsberg, L. (2010). Producing children's corporeal privacy: Ethnographic video recording as material-discursive practice. *Qualitative Research, 10*(2), 249–268.

Angrosino, M. V. (2006). *Doing cultural anthropology: Projects for ethnographic data collection.* Waveland Press.

Atkinson, P., Coffey, A., Delamont, S., Lofland, J., & Lofland, L. (2001). *Handbook of ethnography.* Sage.

Atkinson, P., & Hammersley, M. (1998). Ethnography and participant observation. In N. K. Denzin & Y. S. Lincoln (Eds.), *Strategies of qualitative inquiry* (pp. 248–261). Sage.

Barad, K. (2007). *Meeting the universe halfway: Quantum physics and the entanglement of matter and meaning.* Duke University Press.

Australian Government National Health and Medical Research Council [NHMRC], & Australian Research Council and Universities Australia. (2023). *The national statement on ethical conduct in human research.*

Barad, K. (2010). Quantum entanglements and hauntological relations of inheritance: Dis/continuities, spacetime enfoldings, and justice-to-come. *Derrida Today, 3*(2), 240–268.

Barad, K. (2012). "Matter feels, converses, suffers, desires, yearns and remembers": Interview with Karen Barad. In R. Dolphijn & I. van der Tuin (Eds.), *New materialism: Interviews & cartographies* (pp. 48–70). Michigan Publishing.

Barad, K. (2014). Diffracting diffraction: Cutting together-apart. *Parallax, 20*(3), 168–187. https://doi.org/10.1080/13534645.2014.927623

Barad, K. (2017). No small matter: Mushroom clouds, ecologies of nothingness, and strange topologies of spacetimemattering. In A. Lowenhaupt Tsing, N. Bubandt, E. Gan, & H. A. Swanson (Eds.), *Arts of living on a damaged planet: Ghosts and monsters of the Anthropocene* (pp. 103–120). University of Minnesota Press.

Blom, S. M. (2020). Conceptualizing parent(ing) childhoodnature through significant life experience. In A. Cutter-Mackenzie, K. Malone, & E. Barratt Hacking (Eds.), *Research handbook on childhoodnature: Assemblages of childhood and nature research* (pp. 1–26). Springer International Publishing. https://doi.org/10.1007/978-3-319-51949-4_127-1

Bozalek, V., & Zembylas, M. (2017). Diffraction or reflection? Sketching the contours of two methodologies in educational research. *International Journal of Qualitative Studies in Education, 30*(2), 111–127.

Brewer, J. (2000). *Ethnography.* McGraw-Hill Education.

Brown, S. L., Siegel, L., & Blom, S. M. (2020). Entanglements of matter and meaning: The importance of the philosophy of Karen Barad for environmental education. *Australian Journal of Environmental Education, 36*(3), 1–15.

Creswell, J. W. (2018). *Educational research: Planning, conducting, and evaluating quantitative and qualitative research.* Pearson Education.

Cutter-Mackenzie, A., Edwards, S., & Quinton, H. W. (2015). Child-framed video research methodologies: Issues, possibilities and challenges for researching with children. *Children's Geographies, 13*(3), 343–356. https://doi.org/10.1080/14733285.2013.848598

Cutter-Mackenzie-Knowles, A., Malone, K., & Barratt Hacking, E. (2020). *Research handbook on childhoodnature: Assemblages of childhood and nature research* (A. Cutter-Mackenzie, K. Malone, & E. Barratt Hacking, Eds.). Springer International Publishing.

Day, C. (1998). Working with the different selves of teachers: Beyond comfortable collaboration. *Educational Action Research, 6*(2), 255–275.

Denzin, N. K., & Lincoln, Y. S. (2011). *The Sage handbook of qualitative research*. Sage.

Dingwall, R. (1980). Ethics and ethnography. *The Sociological Review, 28*(4), 871–891. https://doi.org/10.1111/j.1467-954X.1980.tb00599.x

Eisner, E. W. (1992). Educational reform and the ecology of schooling. *Teachers College Record, 93*(4), 610–627.

Grauer, K., & Naths, A. (1998). The visual journal in context. *CSEA Journal, 29*(1), 14–19.

Gullion, J. S. (2018). *Diffractive ethnography*. Routledge. https://doi.org/10.4324/9781351044998

Hammersley, M., & Atkinson, P. (2007). *Ethnography: Principles in practice*. Routledge.

Haraway, D. (1992). The promises of monsters: A regenerative politics for inappropriate/d others. *Cultural Studies*, 295–337. https://monoskop.org/images/f/f1/Haraway_Donna_1992_The_Promises_of_Monsters_A_Regenerative_Politics_for_Inappropriated_Others.pdf

Haraway, D. (2003). Situated knowledges: The science question in feminism and the privilege of partial perspective. In Y. S. Lincoln & N. K. Denzin (Eds.), *Turning points in qualitative research: Tying knots in a handkerchief* (Vol. 2003, pp. 21–46). AltaMira Press.

Harris, A. M. (2016). *Video as method*. Oxford University Press.

Hultman, K., & Lenz Taguchi, H. (2010). Challenging anthropocentric analysis of visual data: A relational materialist methodological approach to educational research. *International Journal of Qualitative Studies in Education, 23*(5), 525–542.

Ingold, T. (2014). That's enough about ethnography! *Hau: Journal of Ethnographic Theory, 4*(1), 383–395.

Jackson, J. E. (1990). "Deja Entendu" the liminal qualities of anthropological fieldnotes. *Journal of Contemporary Ethnography, 19*(1), 8–43.

Jackson, A. Y., & Mazzei, L. (2011). *Thinking with theory in qualitative research: Viewing data across multiple perspectives*. Routledge.

Kellehear, A. (1993). *The unobtrusive researcher: A guide to methods.* Allen & Unwin.

Koro-Ljungberg, M. (2015). *Reconceptualizing qualitative research: Methodologies without methodology.* Sage Publications.

La Jevic, L., & Springgay, S. (2008). A/r/tography as an ethics of embodiment: Visual journals in preservice education. *Qualitative Inquiry, 14*(1), 67–89.

Larson, M. L., & Phillips, D. K. (2013). *Searching for methodology: Feminist relational materialism and the teacher-student writing conference.* Faculty Publications.

Lenz Taguchi, H. (2010). *Going beyond the theory/practice divide in early childhood education: Introducing an intra-active pedagogy.* Routledge.

Lenz Taguchi, H. (2012). A diffractive and Deleuzian approach to analysing interview data. *Feminist Theory, 13*(3), 265–281. https://doi.org/10.1177/1464700112456001

Lofland, J., & Lofland, L. H. (2006). *Analyzing social settings.* Wadsworth Publishing Company.

MacLure, M. (2013). Researching without representation? Language and materiality in post-qualitative methodology. *International Journal of Qualitative Studies in Education, 26*(6), 658–667. https://doi.org/10.1080/09518398.2013.788755

Malone, K. (2018). *Children in the Anthropocene: Rethinking sustainability and child friendliness in cities.* Palgrave Macmillan, Springer Nature.

Messenger, H. (2016). Drawing out ideas: Visual journaling as a knowledge creating medium during doctoral research. *Creative Approaches to Research, 9*(1), 129.

Mills, D., & Morton, M. (2013). *Ethnography in education.* Sage.

Morgan, A. (2007). Using video-stimulated recall to understand young children's perceptions of learning in classroom settings. *European Early Childhood Education Research Journal, 15*(2), 213–226. https://doi.org/10.1080/13502930701320933

Morse, J. M. (1991). Subjects, respondents, informants, and participants? *Qualitative Health Research, 1*(4), 403–406. https://doi.org/10.1177/104973239100100401

Moxnes, A., & Osgood, J. (2019). Storying diffractive pedagogy: Reconfiguring groupwork in early childhood teacher education. *Reconceptualizing Educational Research Methodology, 10*(1), 1–13.

Muir, T. (2010). Using *video-stimulated recall as a tool for reflecting on the teaching of mathematics*. Mathematics Education Research Group of Australasia.

Murris, K. (2020). Introduction: Making kin: Postqualitative, new materialist and critical posthumanist research. In *Navigating the postqualitative, new materialist and critical posthumanist terrain across disciplines* (pp. 1–21). Routledge.

Murris, K., & Bozalek, V. (2019). Diffracting diffractive readings of texts as methodology: Some propositions. *Educational Philosophy and Theory, 51*(14), 1504–1517.

Murris, K., & Haynes, J. (2018). *Literacies, literature and learning: Reading classrooms differently*. Routledge.

Musante, K., & DeWalt, B. R. (2010). *Participant observation: A guide for fieldworkers*. Rowman Altamira.

Nguyen, N. T., McFadden, A., Tangen, D., & Beutel, D. (2013). *Video-stimulated recall interviews in qualitative research*. Australian Association for Research in Education.

O'Brien, J. (1993). Action research through stimulated recall. *Research in Science Education, 23*(1), 214–221.

Parker, M. (2007). Ethnography/ethics. *Social Science & Medicine, 65*(11), 2248–2259. https://doi.org/10.1016/j.socscimed.2007.08.003

Pedersen, H. (2013). Follow the judas sheep: Materializing postqualitative methodology in zooethnographic space. *International Journal of Qualitative Studies in Education, 26*(6), 717–731.

Pink, S. (2012a). *Advances in visual methodology*. Sage.

Pink, S. (2012b). The visual in ethnography: Photography, video, cultures and individuals. In J. Hughes (Ed.), *Sage visual methods* (pp. 123–144). Sage.

Pink, S. (2015). *Doing sensory ethnography*. Sage.

Pink, S. (2020). *Doing visual ethnography*. Sage.

Powell, E. (2005). Conceptualising and facilitating active learning: Teachers' video-stimulated reflective dialogues. *Reflective Practice, 6*(3), 407–418.

Reybold, L. E., Lammert, J. D., & Stribling, S. M. (2013). Participant selection as a conscious research method: Thinking forward and the deliberation of "emergent" findings. *Qualitative Research, 13*(6), 699–716.

Rowe, V. C. (2009). Using video-stimulated recall as a basis for interviews: Some experiences from the field. *Music Education Research, 11*(4), 425–437.

Sanders-Bustle, L. (2008). Visual artifact journals as creative and critical springboards for meaning making. *Art Education, 61*(3), 8–14.

Schwartz, D. (1989). Visual ethnography: Using photography in qualitative research. *Qualitative Sociology, 12*(2), 119–154.

Scott Shields, S. (2016). How I learned to swim: The visual journal as a companion to creative inquiry. *International Journal of Education & the Arts, 17*(8), n8.

Spradley, J. P. (2016). *The ethnographic interview.* Waveland Press.

St. Pierre, E. A. (2011). Post-qualitative research: The critique and the coming after. *The Sage Handbook of Qualitative Research, 4,* 611–626.

Taylor, C. A. (2013). Objects, bodies and space: Gender and embodied practices of mattering in the classroom. *Gender and Education, 25*(6), 688–703. https://doi.org/10.1080/09540253.2013.834864

Thrift, N. (2008). *Non-representational theory: Space, politics, affect.* Routledge.

Verlie, B. (2018). Affective entanglements: learning to live-with climate change [PhD diss., Monash University].

Verlie, B., & Blom, S. M. (2021). Education in a changing climate: Reconceptualising school and classroom climate through the fiery atmos-fears of Australia's black summer. *Children's Geographies,* 1–15. https://doi.org/10.1080/14733285.2021.1948504

Vladimirova, A., & Rautio, P. (2020). Unplanning research with a curious practice methodology: Emergence of childrenforest in the context of Finland. *Research handbook on childhoodnature: Assemblages of childhood and nature research* (pp. 1–26). Springer International Publishing.

Wolfinger, N. H. (2002). On writing fieldnotes: Collection strategies and background expectancies. *Qualitative Research, 2*(1), 85–93.

Woolgar, S. (1988). *Knowledge and reflexivity: New frontiers in the sociology of knowledge.* Sage.

7

THE SEVENTH TURN

Exploring the materialities of context in posthuman educational research

Introduction

This Turn is an exploration of the context of the research including the materialities of place, systems, and world events. As described in the previous Turns, the research presented in this book is situated in the posthuman to explore teachers' perceptions of nature and how it informs pedagogy. When posthuman thinking is adopted, the research focus broadens from being only on the human subjects to also consider the materialities of context such as place, systems and world events. This Turn is an intra-lude, a shorter Turn to explore these materialities that inextricably influence and become the data.

In this Turn, the geographical location of the study – the Country – where this was conducted is acknowledged and introduced. This includes an acknowledgement and overview of the histories and cultures of the First Nations Peoples from Bundjalung and Gumbaynggir Country. I then move to outline the context of the New South Wales (NSW) (see Figure 7.1) public school system in the current political situation as the site of study. The age, context

DOI: 10.4324/9781032703473-7

FIGURE 7.1   The Northern NSW region where this study takes place. *Reproduced from Open Street Maps under a creative-commons licence.*

and socioeconomic aspects of the schools and classrooms are presented at a high level with the detailed context explained at the start of the data Turn – the Eighth and Ninth Turns. I explore the context of the teacher participant, Caitlyn, and myself as the researcher participant to provide an overview of the humans that participated in this study, along with the context of the lessons that informed the lesson participations. Finally, I contextualise this study on a global scale, including world and environmental events that created the backdrop underpinning this study.

Country (geographic location)

I invited teachers working with kindergarten (the first year of formal schooling in NSW, Australia), grade 1 and grade 2 students in public schools in Northern NSW in 2020 to be involved in this research. This region was selected as it is where I currently reside and have the most lived experience with/in. Teachers from the early school years were selected as it was identified that there is a dearth of research in environmental education with early school years' teachers (Payne, 2018). Moreover, these years of schooling are where a lot of students' ideas are formed. Hill et al. (2014) argue that "sustainability education must start early and positively in the lives of young children, before unsustainable patterns of thinking and acting are accepted as the everyday norm and become deeply ingrained habits" (p. 21). The school that Caitlyn worked in is based in the Northern Rivers Region of New South Wales (NSW) Australia (see Figure 7.1). With ethical agency, the exact school title and location will not be disclosed.

This region of Northern NSW belongs to and spans two nations of First Nations Peoples (see Figure 7.2). The Bundjalung Nation in the North has more than 13 different tribes or dialect groups, with each containing several clans or extended family groups (Lismore City Council, n.d.). The Gumbaynggir Nation in the South, which similarly has a wealth of different dialect groups, shares some of their language, traditions, and culture with neighbouring nations. Because of this trait, they were known as the "sharing people" as their land had such abundance that food and other resources were shared with other nations (Arrawarra Sharing Culture, 2009). It is noted that in some depictions, a third nation is acknowledged as being located in between these two nations as a result of Native Land Claim and is known as Yaegl country. This region is bounded by Yamba in the North, Ulmarra in the West, Wooli to the South, and the Pacific Ocean to the East.

First Nations Peoples' histories and cultures

Colonisation occurred in this area in the 1840s under horrific and traumatic circumstances. This led to the displacement of First

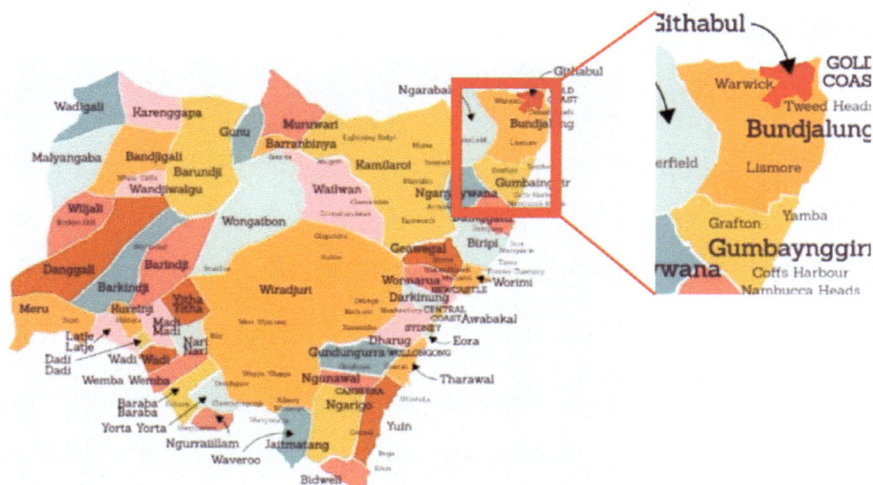

FIGURE 7.2   The First Nations Country where this study takes place. ***Aboriginal Languages and Nations in NSW & ACT*** © ***Reconciliation NSW Note***: *This map is based on the AIATSIS map of Indigenous Australia, which was produced for a general reading audience. The map is not definitive and is not the only information available which maps language and social groups. The information on which the map is based is contested and may not be agreed to by some traditional custodians. The borders between groups are purposefully represented as slightly blurred. They do not claim to be exact. This illustration is reproduced with permission from Reconciliation NSW (www.reconciliationnsw.org.au).*

Nations peoples from their land and families and many were killed under brutal circumstances such as the mass killing at Red Rock, known as "Blood Rock" by the traditional owners (Arrawarra Sharing Culture, 2009; Lismore City Council, n.d.).

Currently, there are still many First Nations peoples living in the Northern Rivers of NSW or the Bundjalung and Gumbaynggir Nations. The schools involved in the study had First Nations students in attendance, and all the students are educated about First

Nations culture, often by a local elder. The teacher participant, Caitlyn, identifies as an Aboriginal Woman from Yamatji Country living now on Bunjalung Country after a tumultuous childhood of displacement. More with Caitlyn is explored in the Eighth and Ninth Turns.

NSW Department of Education: the history and the current situation

At a similar time to colonisation, public schools were launched in NSW. This was in response to the failing of the Church education system which had inadequately educated the entire school-age population; this was most likely due to a reduction in accessible funds during the depression of the 1840s. In 1848, the first NSW government school was established following the model of the Irish National System. Over the following three years, a further 36 public schools became operational, however, only in regional areas where children had no access to denominational schools; there was no intent to offer multiple schooling options for families at this point. Since then, there have been over 30 models of schools trialled in the state, with 15 still in place today. An example of a past offering was the exclusively Aboriginal Schools which were operated by unqualified staff and ceased operation in 1968. An example of a successful model is the Environmental Education schools which were established in New South Wales in 1971 and continue operating to this day (NSW Government, n.d.).

The current public school system in NSW, as with other states in Australia and locations across the world, is largely based on a model that reports to standardised tests (such as the NAPLAN testing system introduced in Australia in 2010) to inform the "success" of a school, its teachers and students. As commented by Prof. David

Kirp (2014), schools have been adopting business models based on competition and disruptive innovation which are both "impersonal" methods and have not delivered the solutions needed. Australia's failing test results and scores have been heavily reported (see, e.g. Zyngier et al., 2012) and highlight how current approaches have not been wholly effective in educating and developing our children. However, current media has been focusing on equality for students in educational opportunities in Australian schools, pinpointing the segregation created by special-entry and private schools; this year also saw a focus on the educational issues associated with COVID. However, despite the shift in media focus and coverage, the issue of low test scores still remains according to the latest PISA results (Thomson, 2019; Thomson et al., 2019).

In response to such issues, Kirp argues that schools need "a champion, someone who believes in them and that's where teachers enter the picture" (Kirp, 2014, para. 2). Despite these comments and observations, schools are mandated to include evidence-based, standardised testing and rewards-based behaviour management systems. However, Kirp and other scholars (such as Nielsen, 2010) advocate for the relational model of education where people are considered first and foremost: their emotional, social, physical and mental well-being along with developing their knowledge and ability to be able to succeed in the world. Moreover, the most innovative and cutting-edge theories and approaches to schooling are not based solely on human-centred learning, but include and incorporate nonhuman nature and the environment as an intrinsic part of the school philosophy and learning programme (Jickling et al., 2018; Malone, 2016; Osborn et al., 2020; Young & Bone, 2020), while also including the human body as an often overlooked part of nature (Blom, 2020; Blom & Crinall, 2020; Weston, 2004).

Given the current media focus on the inequalities in the education system based on socioeconomic status and the difference between private and public sectors, some findings are relevant to the early childhood setting in which this research took place. According to a study by Lamb et al. (2015), four milestones act as an index of opportunity. The milestone of interest in this study is the proportion of children who are "school-ready" across the five domains of physical health and wellbeing, social competence, emotional maturity, language and cognitive skills, and communication skills. The results of the study demonstrate that 78% of the 241, 285 students were considered "school-ready" across the domains, with the remaining 22% deemed not-ready. In the Northern Rivers region specifically, results illustrate that the number of children who are on track across the region is similar to those Australia-wide. However, the proportion of developmentally vulnerable children is significantly higher in the Lismore local government area (LGA) than the average for the Northern Rivers and more broadly, Australia (Seaton et al., 2013). The findings indicate that socio-economic status has the greatest impact, followed by children's attendance in preschool programmes and the quality of the preschool education programme. The impacts of school readiness are shown to be "a pattern of social segregation that persists throughout Australian schooling" (Lamb et al., 2015, p. 14). As this research becomes the top priority for political and media attention, greater pressure is placed on teachers to "reduce the gaps" created by socioeconomic differences and to create better preschool educational experiences. With every new government agenda, fuelled by the neoliberalist structures that the education systems (still) currently operate under, comes an added "thing" that teachers are aware of and which exerts a pressure or a force. Such imposts on teachers and their classrooms are described by Barad

(2007) as material-discursive forces and explain the undercurrent of social, economic, and ethical pressures that teachers face. These considerations provide the social and political context of the education system in operation during this research.

Socioeconomic and cultural contexts of the NSW Northern Rivers region

The Northern Rivers region of NSW is characterised by several key factors including (Seaton et al., 2013):

- A slightly higher population growth rate compared to the rest of NSW;
- A median age of approximately 45 compared to 38 in the rest of NSW;
- A lower median household income compared with NSW; and
- A lower percentage of the labour force that works full-time compared to the rest of NSW.

In the Australian Bureau of Statistics Index of Relative Socioeconomic Advantage and Disadvantage where the top ten areas in NSW in each category were identified, three of the ten disadvantaged regions fall into the area of this study while no areas were considered advantaged (Australian Bureau of Statistics [ABS], 2008). As a measure of engagement in education, the region saw a significantly lower rate of school leavers progressing from year 12 to university compared to the NSW average; in one LGA, where two schools from my study are located, this was six times lower than the state average (Seaton et al., 2013). Census data collected in 2016 (home.id, 2016) reveals a trend for the region compared to greater NSW where more people in the Lismore region have studied less at the bachelor or higher degree level but more at the certificate level. There is a significantly smaller number of people born overseas or who speak a language other than English at home. There are generally fewer full-time but more part-time workers compared with NSW more broadly.

Data entanglements in 2020

As the calendar turned to 2020, Australia was ablaze. Frightening scenes dominated news and social media outlets documenting the very real threat faced by many Australians (Verlie & Blom, 2021). Reports on the exact size of the burnt area vary from 12.8 to 17 million hectares (Werner & Lyons, 2020). To put this in perspective, Tasmania is 6.8 million hectares in area and the Netherlands is 4.15 million hectares. In the close to the six-month duration of the bushfires (Sep. 2019–Feb. 2020) it is estimated that 434 million tonnes of carbon dioxide (CO_2) was emitted into the atmosphere; about three-quarters of the amount emitted by industry in Australia in the previous year (Werner & Lyons, 2020). The loss of life was hugely devastating. Thirty-three humans were killed by the fires and in a conservative estimate, about 1 billion animals were also killed (Werner & Lyons, 2020).

In Northern NSW, the area burnt by the Black Summer fires represented approximately 21% of the total area burned in NSW during the fire season (approximately 1.1 million hectares). Over the six-month period, there were 15 declared bushfire emergencies in the region and approximately 700 homes were lost (Australian Institute for Disaster Resilience [AIDR], 2021). That year, NSW had the lowest rainfall on record and had been in an extended drought, in combination with higher-than-average temperatures – a long, hot, dry summer. In these conditions several dry lightning storms occurred, igniting most of the fires in the region (Australian Institute for Disaster Resilience [AIDR], 2021).

By March, the bushfires were no longer a threat as the season cooled and the major fires were extinguished. At this point, media attention in Australia and across the world turned towards COVID. At this time, the first concerns around economic stability were reported. By the end of March, the news was awash with COVID as, what was commonly referred to as "the virus that stopped the world", took hold. I was making the diffractive data entanglements having just completed the first round of lesson participations. However, as working-from-home restrictions became enforced and NSW schools went into lockdown, my data engagements with the human participants required rethinking. Face-to-face lesson participations were no longer an option as most students became home-schooled unless their parents worked in essential services. Teachers worked on a roster system to teach the handful of students that remained in most schools. No guests were allowed onto NSW school grounds, so I put further lesson participations on hold while anxiously watching the global pandemic unfold. By August, NSW school students had returned to school with restrictions imposed on social distancing and increased sanitisation measures. School visits by guests were permitted in some situations which was at the discretion of the principal. Fortunately, at this point, the research with Caitlyn was approved to proceed.

Conclusion

In this section, I have detailed the spacetime context of this research through exploring Country, educational governance, and local and global happenings in the year 2020. I acknowledged that the research took place on the Bundjalung and Gumbaynggir Nations belonging to the First Nations Peoples of these lands. Colonisation occurred in the 1840s under horrific circumstances, which is also when public schools were launched in NSW.

Currently, schools operate under the standardised testing model despite the criticisms. The Northern NSW region where this study takes place is considered to be somewhat disadvantaged compared to other regions in Australia. The context of 2020 outlined the environmental catastrophes of the Black Summer fires that preluded to this research being undertaken. In addition, I highlighted the impact of COVID on the research through school lockdowns and closures. I argue that these contexts were all crucial as material-discursive practices that inherently informed the research in explicit, implicit and often, unknown ways. The materiality of socioeconomic, sociocultural and historical practices is never isolated, but always already entangled with the here-now and there-then.

The material context acknowledged the Bundjalung and Gumbaynggir Country where this research took place. The history of the NSW Department of Education was explored along with current sociocultural trends of the area that highlight a relatively high population growth rate, a lower median income and a lower percentage of the labour force that works full-time. The spatial situation is then moved to the temporal situation by looking to the local and global context of 2020 through the impacts of COVID and the bushfires that included school closures and lockdowns, and the school setting along with the happenings of the year. I acknowledged the material-discursive forces and affective atmospheres of these entangled seen and unseen happenings on the research.

References

Arrawarra Sharing Culture. (2009). *Fact sheets.* Environmental Trust. http://www.arrawarraculture.com.au/fact_sheets/index.html

Australian Bureau of Statistics [ABS]. (2008). *Index of relative socioeconomic advantage and disadvantage.* https://www.abs.gov.au/ausstats/

abs@.nsf/mediareleasesbyReleaseDate/AC5B967F97D4902E CA257B3B001AF670

Australian Institute for Disaster Resilience [AIDR]. (2021). *Major incidents report 2020–21*. Australian Institute for Disaster Resilience. https://knowledge.aidr.org.au/media/8975/aidr_major-incidents-report_2020-21.pdf

Barad, K. (2007). *Meeting the universe halfway: Quantum physics and the entanglement of matter and meaning*. Duke University Press.

Blom, S. M. (2020). Conceptualizing parent(ing) childhoodnature through significant life experience. In A. Cutter-Mackenzie, K. Malone, & E. Barratt Hacking (Eds.), *Research handbook on childhoodnature: Assemblages of childhood and nature research* (pp. 1–26). Springer International Publishing. https://doi.org/10.1007/978-3-319-51949-4_127-1

Blom, S. M., & Crinall, S. (2020). Growing communities in a garden undone: Worldly justice, withinness and women. *Genealogy, 4*(2), 42.

Hill, A., Nailon, D., Getenet, S., McCrea, N., Emery, S., Dyment, J., & Davis, J. M. (2014). Exploring how adults who work with young children conceptualise sustainability and describe their practice initiatives. *Australasian Journal of Early Childhood, 39*(3), 14–22.

home.id. (2016). *Lismore City local workers key statistics*. home.id. https://home.id.com.au/demographic-resources/australian-census-information/australian-census-data/

Jickling, B., Blenkinsop, S., Morse, M., & Jensen, A. (2018). Wild pedagogies: Six initial touchstones for early childhood environmental educators. *Australian Journal of Environmental Education, 34*(2), 159–171.

Kirp, D. L. (2014). Teaching is not a business. *New York Times*. https://www.nytimes.com/2014/08/17/opinion/sunday/teaching-is-not-a-business.html

Lamb, S., Jackson, J., Walstab, A., & Huo, S. (2015). *Educational opportunity in Australia 2015: Who succeeds and who misses out*. Centre for International Research on Education Systems, Victoria University, for the Mitchell Institute, Melbourne: Mitchell Institute.

Lismore City Council. (n.d.). *The history of Lismore*. Lismore City Council. Retrieved, January 30, 2021, from https://www.lismore.nsw.gov.au/cp_themes/default/page.asp?p=DOC-GBX-81-65-46

Malone, K. (2016). Reconsidering children's encounters with nature and place using posthumanism. *Australian Journal of Environmental Education, 32*(1), 1–15. https://doi.org/10.1017/aee.2015.48

Nielsen, T. W. (2010). Towards pedagogy of giving for wellbeing and social engagement. *International research handbook on values education and student wellbeing* (pp. 617–630). Springer.

NSW Government. (n.d.). *Glossary of school types.* NSW Government. Retrieved, January 30, 2021, from https://education.nsw.gov.au/about-us/our-people-and-structure/history-of-government-schools/school-database-search/glossary

Osborn, M., Blom, S., Quinton, H. W., & Aguayo, C. (2020). De-imagining and reinvigorating learning with/in/as/for community, through self, other and place. *Touchstones for deterritorializing socioecological learning* (pp. 189–230). Springer.

Payne, P. G. (2018). Early years education in the Anthropocene: An ecophenomenology of children's experience. *International handbook of early childhood education* (pp. 117–162). Springer.

Seaton, J., Webb, G., Luckle, K., Evans, T., & Vosz, M. (2013). *Northern Rivers social profile.* R. D. Australia. https://socialfutures.org.au/wp-content/uploads/2015/11/Social-Profile-Update-Nov-2013_Web.pdf

Thomson, S. (2019). Aussie students are a year behind students 10 years ago in science, maths and reading. *The Conversation.* https://theconversation.com/aussie-students-are-a-year-behind-students-10-years-ago-in-science-maths-and-reading-127013

Thomson, S., De Bortoli, L., Underwood, C., & Schmid, M. (2019). *PISA 2018: Reporting Australia's results. Vol. 1. Student performance.* https://research.acer.edu.au/ozpisa/35

Verlie, B., & Blom, S. M. (2021). Education in a changing climate: Reconceptualising school and classroom climate through the fiery atmos-fears of Australia's black summer. *Children's Geographies,* 1–15. https://doi.org/10.1080/14733285.2021.1948504

Werner, J., & Lyons, S. (2020, March 5). The size of Australia's bushfire crisis captured in five big numbers. *ABC News.* https://www.abc.net.au/news/science/2020-03-05/bushfire-crisis-five-big-numbers/12007716

Weston, A. (2004). What if teaching went wild? *Canadian Journal of Environmental Education, 9*(1), 31–46.

Young, T., & Bone, J. (2020). Troubling intersections of childhood/animals/education: Narratives of love, life, and death. In A. Cutter-Mackenzie, K. Malone, & E. Barratt-Hacking (Eds.), *Research handbook on childhoodnature*. Springer International Publishing.

Zyngier, D., Thompson, G., Ewing, R., Eacott, S., & Prasser, S. (2012). New reports sound alarm on school performance: Experts respond. *The Conversation.* https://theconversation.com/new-reports-sound-alarm-on-school-performance-experts-respond-11298

8

THE EIGHTH TURN

Diffractive data entanglements for presenting posthuman educational research – part one

Introduction

I find I'm at my best,
or when I reflect and I feel like I'm their best teacher
is when I'm in an organic mode…
Yeah, you've got to stick to your structure,
your curriculum,
your criteria…
you've got to tick those boxes,
but I find when it's flowing,
or then tomorrow it's like yeah let's do this,
scrap what I planned
we're going to move on to something else.
I can still tick the syllabus boxes
and you can still be learning the intended outcomes,
but let's take it in a different direction.
That's when I feel like they're having the best time,
they're learning more,

DOI: 10.4324/9781032703473-8

and I just feel like a much better teacher
when it's an organic process.[1]

I open this Turn with an artwork (Figure 8.1) and a vignette from Caitlyn[2] as an invitation to engage with the diffractive data entanglements of the research presented in this book. As this book explores early school years teachers' perceptions of nature through posthuman thinking, in this Turn, we turn to Caitlyn's story. Caitlyn is a 33-year-old woman in her fifth year of teaching who identifies as Aboriginal Australian, Yamatji from Wajarri Country. Caitlyn admits that her culture is a motivation to ensure the school is inclusive and culturally appropriate. Caitlyn sees the authentic embedding of culture in the school community as her responsibility and purpose as a teacher because she never received cultural appropriation when she was at school. Furthermore, she is also aware that other teachers do not feel confident to do this, and because she does have a connection to Country, she wants to share this with her school community.

Caitlyn works in a school located in the Northern Rivers of New South Wales and had a population of approximately 34 students in 2020, which changes from year to year. There is a teaching principal and three classes in total, and Caitlyn is teaching the combined kindergarten, grade 1 and grade 2 class.

This Turn explores the diffractive data entanglements (the data diffracted through the diffraction gratings of childhoodnature, affective atmospheres and material-discursive practices [see the Fourth Turn]) with Caitlyn. The data entanglements presented in this Turn, part one, are the first lesson participation, the first video-stimulated recall conversation, photographs of the lesson participation and school grounds, and our visual-journal entries.

FIGURE 8.1 Caitlyn's visual-journal entry: Symbolising the love of being connected to beautiful things in nature. *Original artwork by research participant "Caitlyn". Reproduced with permission.*

It is an entangled and creative journey where images form part of the data and where video-stimulated recall conversation data and certain direct quotes on theory are presented as prose (inspired by the works of Crinall, 2017) to give the data a sense of flow in its becoming. The work is presented in the present tense and active voice to bring you, the reader, into the room with me. To facilitate the flow, references to Caitlyn's quotes are purposefully added as footnotes instead of in-text. These literary acts highlight that the words on the page are "not inert, passive objects in the service, beautification and edification of humanity [in the relationship with the non-human world] … but the lively 'stuff' of matter that is articulate and agential" (Davies, 2021, p. 57). I invite you, the reader, to undertake your own intra-active journey of data entanglements with these data as you read.

Enacting diffractive data entanglements follows the first three steps proposed in the Fifth Turn:

1. On completion of phases 1 and 2 of the data entanglement process (see the Sixth Turn), I re-experience the documented data through a deep dive into watching the video recording of the lesson participation (data from phase 1), looking at the photos from the lesson participation (data from phase 1), listening to the audio-recordings of video-stimulated recall conversation (data from phase 2), and entangling the teachers and my own visual-journal (data from phase 3). While re-engaging with data, I attune to the concepts of childhoodnature, affective atmospheres and material-discursive practices to explore how these concepts grow and expand the data in "new" and unexpected ways.

2. While re-engaging in the data, further field notes and visual-journal entries document how the data's agency emerges – how

the data speaks and makes itself known and in doing so, how further data is generated.

3. The data with each participant is presented individually through patterning from the theoretical diffraction gratings (childhood-nature, affective atmospheres and material-discursive practices).

From this process, eleven diffractive data entanglements became known. They are presented across two Turns – differentiated by different locations in time, and therefore, space (Barad, 2007). In this Turn, we present the first six diffractive data entanglements followed by five in the Ninth Turn. These diffractive data entanglements make lively the ReTurning Learning theoretical framework by giving voice to childhoodnature, affective atmospheres and material-discursive practices through the enactments of Caitlyn, a classroom teacher, in an early school years classroom in our two lesson participations and video-recalled conversations.

Diffractive data entanglement #1: childhoodnature from a withinness

Caitlyn's classroom is a typical classroom set-up with students' tables set up in groups, a designated floor space up the front marked by an orange and red floral floor map, a large television monitor, a whiteboard, student work displays, books, educational posters and a teacher chair (see Figure 8.2). This lesson begins with some group time on the floor mat (see Figure 8.3). Students pack up the activity they are doing, and Caitlyn thanks them sincerely for doing it so quickly. The students are then asked to sit in a circle on the mat.

Caitlyn joins the students by sitting down on the floor in their circle where she does not command a priority position (see Figure 8.3). Although this movement is seemingly unintentional and may not

FIGURE 8.2   A collage of the features of Caitlyn's classroom. *Photographs taken and arranged by Simone Blom.*

FIGURE 8.3 The morning check-in circle. *Photograph by Simone Blom.*

have been something that Caitlyn considers important, these movements, such as sitting on the floor with the students and the way she uses her voice, demonstrate Caitlyn's relationship with, and perception, of nature. Caitlyn holds equal respect for every human body-as-nature, including the intention behind their morning check-in circle (see Figure 8.3) as *"everybody's equal, everybody's relaxed, everyone's here, everyone's present, everyone has a turn"*.

Caitlyn shares with the students that they are going to go outside soon, but first, she holds a discussion about their current topic, Technology in History. Caitlyn gently reminds students about the values that are expected in the classroom when having a conversation. Caitlyn invites the students' input. One of the students contributes that only one person talks at a time. A few other students share what they remember from the last lesson about technology such as that a rock was made into an axe. The students work to have a conversation authentically rather than putting their hands-up. Caitlyn adds direction as needed by providing an opportunity for students who have not spoken to contribute by asking them explicit questions and making it clear to the rest of the group who is speaking to avoid interruptions. The students are evidently eager to be involved in the conversation and are attentively and respectfully listening to Caitlyn as she guides them in their learning. This conversational discussion is reminiscent of the Philosophy of Childhood positioning where children are given authentic opportunities to discuss meaningful matters that are of concern and importance to them (Gregory, 2011). However, the Philosophy of Childhood also advocates for problematising truth which was not explicitly undertaken on this occasion, most likely because Caitlyn was not intentionally attempting to engage in this practice (Murris, 2016).

Caitlyn shares:

I think they're pretty engaged
But it feels to me like they're waiting
they're like okay let's get the talking done
what are we doing next?
Especially this couple,
they're like alright we've said our bit,
we don't want to listen to everybody else,
we've contributed to the conversation
we're ready to move.[3]

In this vignette, Caitlyn describes her reading of the situation – an interpretation of the material-discourses she is sensing through her human body. As Dernikos et al. (2020) explain, "affect is not what you feel, as much as it is an event that forces you to *be(come) affected*, to feel *some-thing*" (p. 5; emphasis in original).

Caitlyn is affected by the material-discursive force of the students' movements and readiness. However, Caitlyn remains steady and encourages respectful listening of each other until everyone has had the opportunity to share something if they want to.

In our post-lesson participation, video-stimulated recall conversation, we turn to the rewards-based system that much of schooling practices are founded upon in current times; a hangover of behaviourist theory. Caitlyn states that *"it is really tricky…[for] some kids that reward [i.e. verbal praise] will be enough for the rest of the session, and [for] other kids you've got to do it continually, at regular increments to make them stay on track"*. She furthers by stating that the rewards system has been different depending on the school and that perhaps, it is the demographic of intrinsic versus extrinsic motivation:

> *When I first got here last year*
> *external motivation drove everything.*
> *So I spent a lot of time trying to build that, like*
> *"are you proud of yourself?*
> *Have you worked really hard for you?"*
> *It's not my learning,*
> *and putting that ownership back onto the kids.*
> *There's no real right or wrong answer yet is there,*
> *it's just this trial and error all the time.*[4]

While just a brief point in the lesson – a single time stamp where some of the kids are getting restless in the group conversation – our conversation brings to the surface that what is happening signifies a dichotomy that is larger than a single moment: that of intrinsic and extrinsic motivation. Through this act of conversation, time is turned upside down, and the small moments do not symbolise the magnitude of the enactment that is performed and observed in each classroom moment. To borrow from a quantum theory position, the micro point of time expanded in space to the macro of contexts and reasoning (Barad, 2007); dichotomous thinking in a multitude of directions. Rather than seeking the externalised reward, the extrinsic motivation, Caitlyn identifies that to encourage her students' agency and responsibility, developing intrinsic motivation is needed.

Perhaps intrinsic motivation is about learning to come together from a withinness (Blom & Crinall, 2020), where "thinking community beyond 'out-there' to also be 'in-here'; where the 'in-here' becomes what 'matters' (Blom, 2020; Kahn & Hasbach, 2013; Payne, 1997)" (Blom & Crinall, 2020, p. 15). This thinking that has been highlighted by Caitlyn demonstrates her way of thinking with/as nature, a childhoodnature moment, where the withinness and knowing from within the body as nature is brought into awareness, acknowledged, and valued (Blom, 2020). Perhaps this is Caitlyn's way of knowing nature through being responsible and from a withinness, "in here". Furthermore, this moment in time, that is returned to and turned over and over again, as described by Barad (2014) through the analogy of an earthworm's movements, explores the mattering of existence. This point in spacetime is diffracted out in infinite directions, one of which has been highlighted here.

Diffractive data entanglement #2: frogs, ants, dragonflies, and bee-ing childhoodnature

The day is hot with a maximum temperature of 34.5°C. As such, Caitlyn time outside to 15 minutes outside for students to collect their materials, before returning inside to make their technologies; this was part of her duty of care and to ensure no one got too hot. Each student receives a container to collect their materials and as the students begin getting up to go outside, Caitlyn states that the students can work either individually or in a pair. Caitlyn further reminds the students to get a hat because it is a hot, sunny day. Outside, Caitlyn's movements are fluid, focused yet extremely aware of all the students in her class. Although the students are spread across the entire schoolyard, I get a sense that she is "holding them all"; aware of their every movement. Caitlyn is clearly not anxious about the students running around in the yard, exploring nature. This is not a naturally easy task for teachers, some who find the lack of physical boundaries of the outdoors daunting and this is a large limitation and inhibitor for students to engage authentically with the natural world. Much research has identified teachers' fears and many other reasons for inhibiting these childhoodnature opportunities (Barfod, 2018; Glackin, 2018; Jacobi-Vessels, 2013; McFarland & Laird, 2020; Waite, 2011). I journal on this observation:

The four walls of the classroom
must have been written about millions of times over the
centuries
since school time began.
Moreover, they have been stared at blankly
for many more hours than that
by the millions of children
that have been a part of their story.
The boxiness and boringness of the drab walls,
disguised
by colourful posters, artworks, reminders, teachings.
Perhaps these wall coverings act
more like a distraction than a disguise.
The busyness of classrooms.
The noise covering the not-so-quiet

story
of the bare walls that indeed
have their own story
to tell.
They tell it constantly,
how can they not?
It is their responsibility,
response-ability,
as material-discursive agents in the world.
The walls enact their part
by sharing their story.
Are we reading it?
Are we listening?
Are we enacting our part in this performance by responding?
Respond-ability.[5]

The children excitedly approach Caitlyn as she continues to roam around the yard. They run to her with their findings in a "show and tell" moment between them and her. Caitlyn is genuine in expressing her delight and equal excitement about the treasures that they have found. She comments while connecting deeply with the student by asking them a question or tuning in to their excitement and ideas. In these "moments", Caitlyn continues to move with purpose around the yard. Caitlyn appears at ease outside the classroom, moving fluidly between the student groups. One of the children finds a small frog and picks it up. Caitlyn approaches the student and explains how to gently hold the frog and suggests to the student to find somewhere cool and shady to return the frog to nature.

As I film Caitlyn walking through the yard, I notice a wooden post that has a train of ants walking up and down. I stop momentarily to watch the ants. Their flow of movement, communication and working together reminds me of Caitlyn and her class: the constant communication of nature; a childhoodnature moment where nonhuman nature (the ants) is attuned to the way the children's bodies are moving and working together. The ants move together on the built environment and structure, like the students moving around their built environment of the schoolyard, and Caitlyn – like the Queen – encourages communication from the students. At this moment, the ants speak loudly and become the focus of my attention as they tell their story and communicate through their movement. The material-discursive practice of the ants highlights the way the class is working together and the synergy between the human bodies as nature and the ant bodies are expressed to demonstrate their similarities.

Ants have the adaptability to live in almost any environment and tend to dominate any environment in which they live. Man [sic] is the only other animal species of which this may be said. In addition, the ant is truly a social insect. Of all insects, the ant is the most manlike [sic] in behavior and the most adaptive. Some other insects form inflexible societies, but ants show great variation in behavior and social organization even within the same species. And most interesting is the fact that the patterns of social and biological evolution of ants are parallel. As the physiology of ants evolved there was a parallel evolution in social organization. Since many of the

*4000 species of ants represent evolutionary
atavisms, this pattern of parallel evolution
may be seen, not just reconstructed.
Therefore, study of ant evolution may
provide clues to the study of human evolution…
both physiological and social.*[6]

I consider the human-animal relationality here through the concept of "withlings" proposed by Tammi et al. (2020). Although the ants' class as child-animal relations are not identified by the children themselves, I see the significance of the concept of "withlings" in understanding the "complex and often conflicting emergence of child-animal relations" (Tammi et al., 2020, p. 1310). This posthuman concept espouses childhoodnature through placing,

Emphasis on the shared processes
through which relations take place
rather than on individual (human) views of these relations…
where the environment remains
as a mute context for human activities…
[and] humans and their nonhuman surroundings
do not exist as independent of each other.
Tammi et al. (2020, p. 1312)

Caitlyn continues connecting with the students and roaming around the yard. The sound of cicadas is loud and starts to dominate the video-scape; another moment of nonhuman nature making itself known. Caitlyn is carrying a small spade she has found in the yard with her. While she is talking with a student, she notices a bee on them and uses the spade to confidently and quickly flick the bee away. There is no fuss, no fear, no harm. The bee lands on the squishy "soft fall" floor of the outdoor playground (see Figure 8.4) and the child continues on their search for more materials. Perhaps in this instance, the bee is instigating the child-animal relation as a movement of co-becoming or becoming with as "withling" (Tammi et al., 2020). Caitlyn shares,

> … *It's expected, there are bees;*
> *we have wasps in our classroom all the time…*
> *I have a dragonfly tattoo on my shoulder,*
> *for me, I feel like dragonflies are my totem,*
> *and I talk to the kids about totems and things like that.*
> *So when we get dragonflies,*
> *and last year in the last term we had about five dragonflies stuck*
> * in our classroom,*
> *and they just thought it was just this magical good luck thing*
> *because we had all these dragonflies in here (see Figure 8.5).*
> *It was a really cool moment [Chuckles],*
> *yeah they love it when the nature comes inside.*[7]

FIGURE 8.4 The bee on the playground soft fall. *Photograph by Simone Blom.*

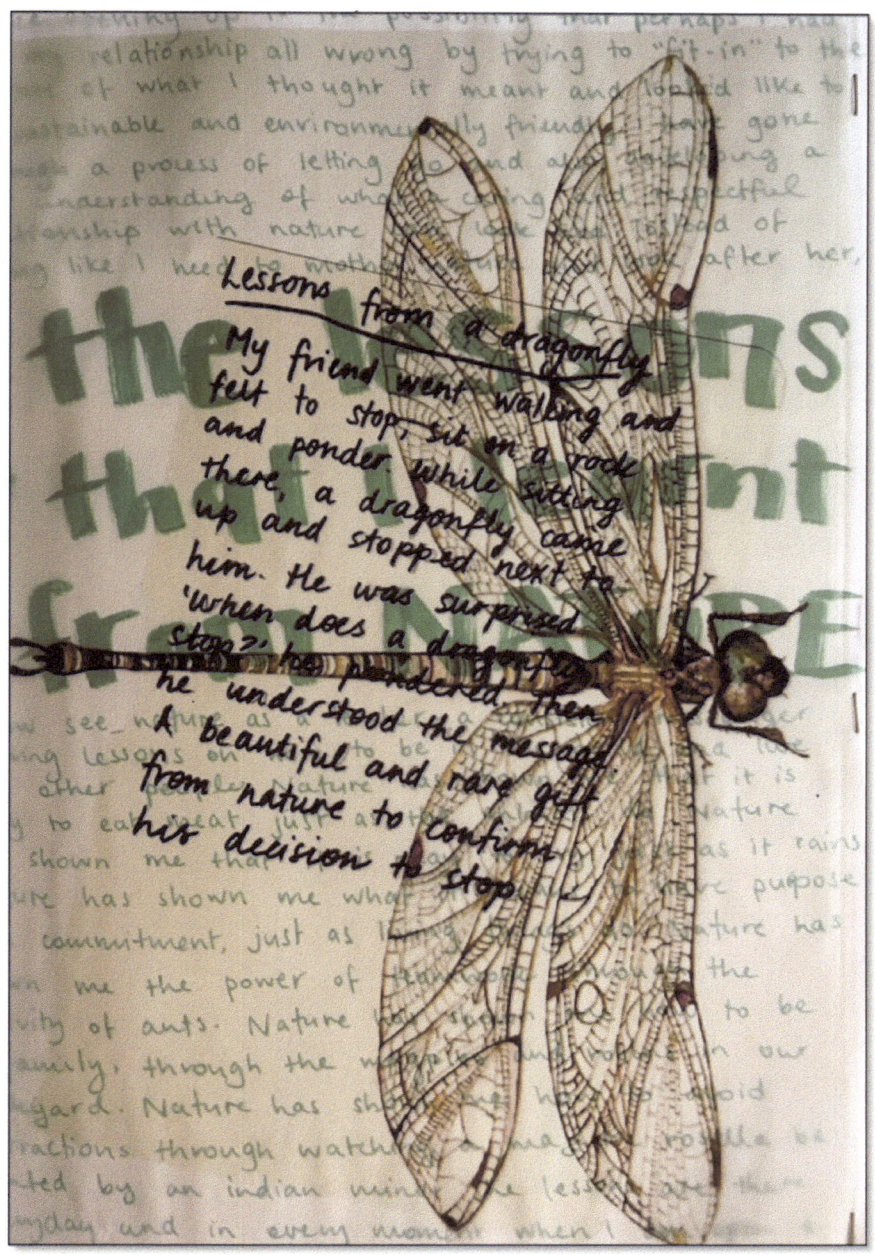

FIGURE 8.5   Simone's visual-journal entry: lessons learnt from nature by a dragonfly. *Original artwork by the author.*

Through these human/nonhuman animal intra-actions, Indigenist theorising of relationality is demonstrated which also aligns with the posthuman position that espouses "the inseparability of humans from their material environment and more-than-human relations and response-abilities" (Murris, 2020, p. 50). Highlighting the animal in these lesson participations are enactments of "intra-active relationality" which include nature and culture and where *"the diffractive teacher can be human, nonhuman or more-than-human"* (Murris, 2020, p. 51; emphasis in original).

I consider how First Australians cultures describe human-animal connections and relationality through the concept of "totems". In Australian Aboriginal culture, having a totem means that you are "responsible for looking after that animal or plant and its habitat", whereas looking after means you "never, ever kill your totem animal" (Wilks, 2014, pp. 3–4). While the human species have been responsible for the extinction of countless living organisms, First Australians cultures placed responsibility on the human to be the custodian of their totem animal or plant species. These human-animal relations between people and totems ensured the sustenance of all species in a relationality of co-becoming with, instead of domination over, to enable the balance that existed in Australia until colonisation.

Diffractive data entanglement #3: time and childhoodnature

After about ten minutes, the students return to the classroom. The room is abuzz, not with bees, but with children who are interested, excited, and focused on their learning – with their nonhuman nature treasures found outside the classroom; a childhoodnature enactment. The children are excited to be making their piece of technology using string, sticks, mud, leaves, and anything else they found in the yard. The children's fine motor skills are being challenged by using the string. Caitlyn is quick to see the problem and responds immediately by sitting on the floor with the students, talking them through the process rather than just doing it for them. Caitlyn talks about "taking their time" and using this to problem solve. The omnipresence of time as a material-discursive agent is again making itself known.

The concept of "taking my time" is one I have continued to find interesting, especially within the context of schooling. Schools are governed by clock-time at the expense of students' own time, resulting in students often feeling rushed to get through what they need to, very much *out* of their natural rhythm; a movement away from childhoodnature. Crinall et al. (2020) question the concept of time, and my time: "When do I *not* have my own time? When we are always in our bodies, how can there ever be time 'away' from self?" (p. 67; emphasis in original). Crinall et al. (2020) trouble the concept of clock time, linear time and school time to leave children, adults, and all people to experience no time; as was done during covid lockdown. This way of conceptualising time is considered through the idea of nature's cycles (Crinall et al., 2020) and Indigenous knowledge that proposes "circular time" (Aikenhead & Ogawa, 2007). I consider that young children might not even know what "take your time" means or how to conceptualise having to understand moments through time-based limitations (see Figure 8.6). Indeed, Murris and Kohan (2021, p. 11) argue that,

> Time cannot be considered
> any longer as chronological
> in the sense of being a line of
> consecutive,

FIGURE 8.6 Simone's visual-journal entry: my time, your time, clock time, clockections. *Original artwork by the author.*

irreversible
and uncontaminated
sequence of discrete movements,
moving according to the numbering
of "before" and "after" relationships.
In spite of this onto-epistemological insight,
Unilinear chronological time
continues to govern our world
and lives.

Later, in our second lesson participation, Caitlyn states *"we've got about fifteen minutes to get a really good start happening here"*. I ponder again, what does time mean to these children? Do they have a sense of time? Perhaps, time is nothing (Crinall et al., 2020). Rooney (2016) troubles the notion of time more broadly in the lives of children through the concept of childhood. Rooney (2016) challenges positions of childhood that are linear accounts of children "becoming", instead alerting the reader to the messy and relational possibilities of childhood that are generative and dynamic, resisting the fixed patterns of highly structured routines and patterns. I consider what this moment would look like without the pressure and limitation of the time-based 15 minutes if children were not given a limit, but allowed to live more in what they are doing, allowing "children's potential [to be seen] as something vital and dynamic" (Rooney, 2016, p. 197).

I consider the actions of the time-based limit on Caitlyn's perceptions later in our video-stimulated recall conversation:

I have noticed myself since coming back [in term 2 after Covid lockdown]
there was such this –

the classroom is so high paced all the time and
we've got to do this,
we've got to do this,
we've got to implement the L3 program which means
you've got to do this.
Like all these things,
visible learning,
you just get thrown all of this stuff to juggle
in this tiny little space
for these little human beings,
you know what I mean?
It's just too much.
And I said to my boss the other day, I said
after all this covid stuff and doing –
I didn't do online learning like digital learning,
it just wouldn't suit the demographic of my families at all
so I just made up learning packs.
I made a science booklet which covers the content that we were
    going to cover anyway
and I just put it together,
and we've been able to bring everything that I sent home
straight back into the classroom
and we're still using it....
we've just made a booklet with all their little bits,
I put Dreaming Stories in there,
and they've just had to do little chunks of the work through the
    booklets.
So I did it more that way and just printed little things and sent
    them all home.
And I just thought it's just nice to pair it back and not have –

I said

"that's what I'm taking out of this now",

I said

"I'm not doing that fast pace madness anymore",

I said

"we're just going to cruise through our day and have

quality over quantity

back in that classroom".

That's the biggest thing I've noticed learning wise, and I think
*   it's really changed*

because I am able to be a bit less reactive.

It's not like stop talking, don't interrupt me, I've got to get this
*   done, or*

we have to finish this part and we have to do this,

it just feels a bit more relaxed and all the kids do a little bit
*   more.*[8]

During term 2 of 2020, I initiated a series of PhD updates with my participants as a point of connection during what is a very disconnecting and strange time for many during and post the covid lockdowns. This piece on time, is taken from one of the newsletters (also see Figure 8.7):

A few years ago, I conducted an autoethnographic study to
explore my perception of nature as a parent.

Time as a concept was represented as a key finding in my results.

It was evident that children were spending less time in nature and
more time on screens due to their time-poor lives (Blom,
2020; Malone, 2007).

At that time, I felt like I too was becoming the victim to time
in that I had experienced

FIGURE 8.7 Simone's visual-journal entry: time-based prison on earth. *Original artwork by the author.*

this phenomenon and observed that
it dictated many of my movements.
In response,
I decided to consider how I conceptualised time to enable me
to do what was needed (i.e. I still needed to get to work on
time and meet deadlines)
while feeling a connection with nature.
Moreover, I considered how I modelled a relationship with
nature
when my busy work and study commitments
took me away from these moments.
The idea of "nature-deficit-disorder" (Louv, 2006) was well-
known around this time and I could see that, despite the
best intentions by the author,
it created a force (material-discursive) that parents should
be spending copious amounts of time outdoors with their
children.
I loved this idea in theory and thought about early pedogogues
such as Froebel or Rousseau
who advocated for childhoods where learning was based
purely in nature.

…time can be reconceptualised through deidentifying with the
linear concept of time and reunderstanding time through the
quality and not quantity of our interactions with/as human and
non-human nature…Adopting a non-linear, spatialised view
allows each body (human and non-human) to interact in the
rhythm, movement and discourse of nature.

Blom (2019)

However, as a working mother, I knew that this idyllic picture was not possible
and decided to work with what I could.
Instead of a nature-deficit-disorder,
I considered the idea of a time-deficit-disorder.
I grappled with this notion and started experimenting with prioritising
how I was being (the quality) rather than how much I was doing (the quantity).
This enabled a stronger focus on relationality and less on temporality. (Murris & Borcherds, 2019;
Murris and Borcherds (2018, p. 162) discuss the perception of time
from the adult and child perspective:

Childhood conceptualized as the period at the start of a person's life presupposes a quantitative, chronological concept of time, thereby imposing an adult form of temporality...or linear and irreversible time. However, childhood is not just a period in a human life, but also a particular relationship with, and experience of time … – as associated with play and designates the intensity of time in human life – a destiny, a duration, …an un-numbered movement, not successive, but intensive – but a particularly forceful and intense experience of being-in-time: childlike.

The concept of relationality, is demonstrated through Indigenous ways of knowing and being as many Indigenous cultures are relational by nature and are attuned with the cycles of nature. The following excerpt explores a comparison of Indigenous knowledge and western minority world science through the concept of time (rectilinear/western time and circular/Indigenous time) and knowing nature:

One alternative to Eurocentric science's rectilinear time is circular time (Peat, 1994), a concept of time that harmonizes with the myriad of cycles observed in nature. The idea of all things being in constant motion or flux leads to a holistic and cyclical view of the world.

Aikenhead & Ogawa (2007, p. 563)

In a similar way, but through the theories of quantum mechanics, Barad (2010) troubles the existence of linear time regularly. In this theorising, time – past, present and future – are taken out of linear limitations and re-framed using atomic theories.

What is needed is an understanding of temporality where the "new" and the "old" might coexist, where one does not triumph by replacing and overcoming the other. Quantum superpositions and, relatedly, quantum entanglements open up possibilities for understanding how the "new" and the "old" – indeed, multiple temporalities – are diffractively threaded through and are inseparable from one another.

Barad (2017, p. 69)

In exploring time through visual-journaling this week,
I consider how much the world is "locked-up" by time
(there is a little clock symbolising the lock in this picture
[Figure 8.7]).
I also wonder how these theoretical ideas are conceptualised
and practised
in everyday situations: how they "look" in practice.
It would make sense that there is a need for some measure of
time,
for some way for time to be quantified to enable communica-
tion and organisation;
perhaps this was why the mechanical clock was invented in
the 14th century.
This in itself is noteworthy: that mechanical time was
"invented" 600 years ago and yet, "Eurocentric sciences
embrace rectilinear time as an absolute feature of reality".
Aikenhead & Ogawa (2007, p. 548)

What would time look like if it was not the reality?[9]
I spent the last three weeks [of term 1 during covid lockdown]
working from home, because we had no students here,
from that point nobody came in anyway.
But I was finding myself
up at 6:00 o'clock in the morning on the computer till 10:00
pm at night,
and constantly at people's beck and call,
because people were like
"can you sort this out for me, do you know how to do this, can
you fix these?"
or blah, blah, blah,

"show me how to do this".
So it was just constantly on and
I was like oh this is really not healthy,
I'm working more than I would have been anyway,
so it's just quite a bizarre time…
it's really imposing if you let it.[10]

Diffractive data entanglement #4: fracturing misopedy, childhoodnature

The way Caitlyn moves around the space with the children authentically respects the children as equal others: from the way she sits with them on the floor to the way she speaks with them as she would to me or anyone else. There is care in what she has to say but it is not loaded with condescension or a "better than" stance. It is a matter of fact and in a way that you know she is looking out for you; Caitlyn has your back no matter your age or anything else. This is also demonstrated during the video-stimulated recall conversation when Caitlyn shares her grappling with the way that parents and teachers are always telling kids to "be careful".

She states that,

We do it like
"don't play with sticks,
don't climb the trees,
don't do this"…
Where is the risk taking and
that connection to being outdoors
and in nature
and just being kids?[11]

These perceived "risks" associated with outdoor play and learning inhibit many teachers from taking their students outside into nonhuman nature; indeed, "in Minority western cultures, children today spend more time watching television and being indoors than they spend being active in outdoor environments" (McFarland & Laird, 2020, p. 1077).

Furthermore, Caitlyn's observation that the "be careful" narrative no longer works for her as a teacher or parent, demonstrates a rejection of the superior "all-knowing" teacher position. In doing so, Caitlyn acknowledges children's agency and knowing, adopting childhoodnature thinking. I consider this through the work of Murris (2020) who attests that "deficit figurations of child (child as not fully-human-yet) and colonialism are entangled phenomena" (p. 16). Caitlyn is resisting the material-discursive forces of typical teacher behaviours, which can include misopedy, often unintentionally. Caitlyn is grounded in a different way of being with other; a relational intra-action that may stem from Caitlyn's First Nations culture that troubles colonist ontologies. Murris (2020) continues to describe how, "like Indigenous peoples, children are seen as simple, non-abstract, immature thinkers who need age-appropriated interventions in order to mature into autonomous fully-human rational beings, and therefore cannot be granted political agency" (p. 17). With/in this context, the relationality between Caitlyn and her students could run much deeper than even she might be aware of. Perhaps, the structures that are established through our "social norms" to unintentionally treat children as lesser beings are equally projected onto First Nations peoples through political correctness, all forms of niceties and forever *trying* to do the right thing; trying to help and make things "better". These intentions (still) position the other as less – like the one making the judgement knows more and is in a superior position to make such an assessment.

I am reminded of an image of Karin Murris (Murris & Borcherds, 2019, p. 200; see Figure 8.8), where Karin purposefully sits on the floor of a university classroom to make herself vulnerable. The vulnerability of the smaller human body of the child. It is humbling to consider that the differences that keep the dichotomy of child/adult alive are founded on the physical size of human bodies, and, as such, the impost of linear time. Murris and Borcherds (2019) remind us that "each person is more than a body, always connected, embedded and embodied, dynamic and active" (p. 205).

FIGURE 8.8   Karin Murris placing her body on the floor in a position of vulnerability. *From childing: A different sense of time by Murris, K., & Borcherds, C. (2019b). In B. D. Hodgins (Ed.), Feminist research for 21st-century childhoods: Common worlds methods (p. 200). Copyright 2019 by Bloomsbury Publishing. Reprinted with permission.*

Murris (2020) considers the perpetuation of misopedy not only through the *"moral* superiority of the adult coloniser" but the incessant use of time as a form of linear progress in schools (p. 48; emphasis in original).

Diffractive data entanglement #5: hats and gender normativity

Caitlyn informs the class that they will have 15 minutes to complete their inventions before presenting them to the class. I consider that the notion of 15 minutes is most likely somewhat meaningless to such young children. Perhaps it offers a way of sensing based on the way Caitlyn communicates. For example, there is an urgency in Caitlyn's voice as she announces that there are about five minutes left to pack up. This urgency alerts the children to the impending ending of the lesson. Many of the students come up to Caitlyn and start enthusiastically sharing their inventions. Caitlyn reminds them to pack up first and sings "hocus pocus" and the class responds, "everybody focus". Caitlyn further sets her phone timer for three minutes. The excited affective atmosphere of the room continues. The children are noisy and some of the younger children are finding the excited classroom atmosphere more challenging to be in than others.

I notice that many of the children are still wearing their school hats inside. I have observed in my daughter that when she wears her school hat when it is not needed, it is like she is hiding. I not only find it difficult to *see* her, but also to *feel* her being herself and with a sense of confidence. It is like she stands taller and speaks more loudly when her hat is off even though I know this is not strictly the case. I consider the material-discursive force of the hat. Its purpose is to hide children from the sun – the harmful effects of nature – but in doing so, could it act to inhibit the child? Perhaps to hide the child from the sun, as nature, the material-discursive force of the hat acts to inhibit childhoodnature. As the children sit down on the mat, Caitlyn reminds them to take off their hats. I wonder why children need to be reminded of this and consider that they are so distracted and perhaps even overwhelmed by everything they are feeling and sensing through their bodies that their hats are forgotten.

Caitlyn asks the students who would like to share with the class, and the first couple of children get up to present their inventions. Caitlyn says she would like to take some photos to send home through ClassDojo[12] and asks if this is okay. Again, this respectful act demonstrates the way that Caitlyn is with the students – ensuring they are treated equally. The students' class presentations begin. Caitlyn scaffolds the presentations by asking the children what their invention is, how they made it, what the purpose is and why they need it. Caitlyn sits on the floor with the rest of the class. The first two students, who are girls, present the jewellery that they made.

Caitlyn later shared in our video-stimulated recall conversation,

I find with this group of kids
there's a stark difference between
cliché boy type nature
and
the girl type nature…
You know not to be stereotypical,
but the girls are quite feminine, they've made jewellery and stuff
*   like that,*
there's that real big spectrum
of real feminine desires and real masculine boy hands-on sort
*   of stuff…*

Sometimes it's really tricky to manage with them all,
but these sorts of activities are really good
because they all get to do what they want,
they can find an avenue to access it.[13]

Caitlyn brings to the surface one of the many underlying dichot-omies that are perpetuated in the classroom: gender identification. In a world that is progressing in a seeming whirlwind of gender transitions, transformations, confusions and undoings, Caitlyn's awareness of how this is seen and experienced in her kindergarten to grade 2 classroom presents a childhoodnature moment where binaries are challenged.

Israel,[14] *the little boy with long hair,*
when we have guest teachers or
we have other people in the classroom they often think he's a girl,
Israel is quite a neutral name
and he's got very gorgeous long hair.
So we've talked about that a few times,
because it happened just the other day,
[a teacher came in] to do a demonstration with something
and the kids all looked to each other
and he looked at me.
And then when the teacher was gone
"I know" I said
"but does it really matter, do you care?",
and he's like "no",
and I was like
"yeah so who cares,
it doesn't matter if they're girl or boy,
we're all just kids aren't we".

So that's the extent of what we talk about,
but we don't really discuss the gender stuff no...
See I don't ever think about it as girl or boy,
more feminine masculine,
but not that feminine means woman or girl,
or masculine means boy or man or anything.
I think about it more as
feminine in light and those particular ways, or
masculine that way instead of a separate gender.[15]

The role of gender has been questioned, raised and contested in many educational settings and in particular, given the context of Caitlyn's lesson, in science and technology education extensively in recent years (for example, see the discussion in Chapter 7 of Wolfe, 2021). In regards to the role of gender in science studies, Barad (2011) critiques traditional social constructivist views in science that purport the nature/culture dualism such that "the gendered assignment of the social in the active (gendered male) role of agent inscribing the passive (gendered female) slate of nature, obscures crucial dimensions of power" (p. 449). Barad (2011) further argues that their engagement with the feminist is not about women or gender but a feminist understanding of the political. I consider how people might associate feminist theorising with being a female or the women's rights movements instead of understanding the feminist in adopting a minority view to engage with the world. Caitlyn acknowledges this,

It's a very thought-provoking conversation to have isn't it,
to think about gender and
children and
what do you do with that and
what are you putting on them?

The "what we are putting on them" is the material-discursive practice of gender that is perpetuated through social constructs and norms. Wolfe (2021) presents the affective discourse that happens in science classrooms around gender through belonging. In particular, "how *being a girl* (assumed as always prior) creates particular, reductive ways of affective belonging at school" (p. 128). These normalised ways of being include power dynamics that "continue to privilege some student bodies (generally white and male) at the expense of all others" (Wolfe, 2021, p. 129). Indeed, Caitlyn's observations on gender provide insight into the happenings in the early school years classroom.

Diffractive data entanglement #6: snakes and ghosts

During the presentations, a child returns from the toilet looking quite traumatised as they say they have seen a snake.

Snakes in summer venture closer to our homes in search of food and water.
Try clearing debris from around the home, and
keeping the grass and hedges under control…
Somewhere between 70 and 90% of people who get bitten by a snake,
get bitten while trying to kill them.
In the Northern Rivers, there are numerous snakes per square kilometre.
Wildlife Information Rescue and Education Service [WIRES] (2021)

Caitlyn invited the child to sit on the other side of her on the floor, next to two other children who appeared concerned and very caring. Caitlyn did not make a fuss or big drama but gently put her hand on the girl's back. Barad (2012) shares, "so much happens in a touch: an infinity of others – other beings, other spaces, other times" (p. 1) and through quantum theorising offers *"Touching is a matter of response. Each of "us" is constituted in response-ability. Each of "us" is constituted as responsible for the other, as being in touch with the other"* (p. 7; emphasis in original). I made a visual-journal entry on snakes (Figure 8.9).

The presentations continued and Caitlyn did not skip a beat in facilitating and scaffolding the learning while the three students finished sharing their technology creations. However, the affective atmosphere of the room has changed, and it appears that Caitlyn is very much aware of it. Caitlyn gives the students directions about how to move their bodies: *"hands in the air, and now move your fingers all the way down to your toes"* and *"hands on hips, sway them from side to side"*. While she is giving directions, Caitlyn asks the child who saw the snake to come with her as she moves to her desk and asks another child to present their technology.

The room becomes still and quiet as the child and Caitlyn begin to have a private conversation a few metres from the mat. The conversation lasts less than 20 seconds, but the child and Caitlyn resolve what is needed, and the class presentations continue. Caitlyn uses her classroom phone to call in some assistance and another adult comes and collects the child from the classroom. The restlessness of the class subsides, and the children are re-engaged and focused on each other's presentations. However, Caitlyn understands the calm atmosphere may be temporary as the children are getting ready to move on to the next activity.

After packing away their technology, they regroup on the floor. This time, Caitlyn sits on the teacher's chair, and they discuss the child who has "seen the snake" as the other children are very

I was sitting in our front room the other day speaking on the phone to a friend about
bad habits and behaviours that we continue to engage in when we know they do
not support us and they throw us off our game. Those things that
we laugh off perhaps as being impossible to change - you can't
teach an old dog new tricks and all of that. We spoke about
how sometimes we really deny their existence, they remain
as a blind spot where we know that the blind spot is
there but we don't actually want to see what is
there. At that moment, a red bellied black
snake came into view through the
window into the front garden. I had
never seen one before and knew
from the timing that nature
was speaking in confirm-
ation of my conversation
It represented the
dark and 'blind' that
we don't want to see...
the deeper understanding
of how our seemingly
harmless behaviours and
actions have an impact on
everything... they are the
'poisin' that infects us all
what's lurking in the grasses?
what's hiding in the bushes?
what are we protecting in our blindspots...
the red-bellied black snake...

FIGURE 8.9 Simone visual-journal entry: The red-bellied black snake. *Original artwork by the author.*

concerned. Caitlyn offers that there may not have been a snake and reminds the children, *"you know sometimes you see something or you're just outside and you hear a noise and then in your imagination you think you've seen something and then you get yourself really worried. I think that's what's happened".*

Furman (2022) writes about welcoming entanglements with ghosts into figurations of childhood. In doing so, Furman (2022) reminds us that childhood and "children known for blurring the boundaries between the real and the imaginary" (p. 255). However, Furman (2022) continues that "the common adult response is the corrective assurance … that visions are unreal" (p. 255). This discussion highlights that these "corrective assurances" can act as "emphasising deficits" that can undermine the "child's capacity as a knower and actor" (Furman, 2022, p. 256; see also Murris, 2016). Acknowledging these acts as confirming the adult/child binary position, reveals the gross extent of "business-as-usual" practices that we can bring to the surface through these diffractions. Theorising with posthumanism brings to light practices that unintentionally silence children – such as not talking about ghosts or snakes – and encourages pedagogies that always listen to others in equalness. Such exposés are uncomfortable as much as they are enlightening to what we do as teachers. This is lifelong learning in practice as adults and teachers and how we can deepen our relationships with children, where

> children are knowers who live amongst the ghosts…
> In taking a posthumanist approach
> to explore what lives in their "bones and sinews"
> (Osgood, forthcoming),
> teachers can be responseable to children

as they tangle with ghosts.
In doing so,
they can help children…
re-member and, in re-membering,
take response-ability.
Furman (2022, p. 262)

Conclusion

This Turn demonstrates how the Returning Learning theoretical framework can be put to work, by detailing the intra-active movements of a classroom teacher, Caitlyn and her intra-actions with students and nonhuman others in an everyday lesson. The diffractive data entanglements generated from the lesson participations, video-stimulated recall conversations, visual-journal data and emails, explored Caitlyn's perception of nature and how this informs her pedagogy.

Through pushing the data through the diffraction gratings of childhoodnature, material-discursive practices and affective atmospheres, perceptions of nature and the informed pedagogy are presented through six diffractive data entanglements. The generative exploration through this Turn highlighted nature from a withinness, multispecies engagement, relationships with time and gender, misopedy and ghosts. The following Turn continues this exploration in a second lesson participation and video-recalled conversation with Caitlyn.

Notes

1 Caitlyn, video-stimulated recall conversation, 19th February 2020.
2 Caitlyn is a pseudonym.
3 Caitlyn, video-stimulated recall conversation, 19th February 2020.
4 Caitlyn, video-stimulated recall conversation, 19th February 2020.

5 Simone's journal entry.

6 Reinhart (1980, p. 100).

7 Caitlyn, video-stimulated recall conversation, 19th February 2020.

8 Caitlyn, video-stimulated recall conversation, 11th June 2020.

9 Simone, Term 2 Week 6 Update, 29th May 2020.

10 Caitlyn, video-stimulated recall conversation, 11th June 2020.

11 Caitlyn, video-stimulated recall conversation, 11th June 2020,

12 ClassDojo is an online, interactive rewards-based system which originated in the United States of America. Promoted as a communication tool, the website claims that 95% of all K-8 schools in the U.S.A. use ClassDojo and that it is used in 180 countries (ClassDojo, 2012).

13 Caitlyn, video-stimulated recall conversation, 19th February 2020.

14 Israel is a pseudonym used to ensure the student is unidentifiable thus maintaining ethical privacy.

15 Caitlyn, video-stimulated recall conversation, 19th February 2020.

References

Aikenhead, G. S., & Ogawa, M. (2007). Indigenous knowledge and science revisited. *Cultural Studies of Science Education, 2*(3), 539–620.

Barad, K. (2007). *Meeting the universe halfway: Quantum physics and the entanglement of matter and meaning.* Duke University Press.

Barad, K. (2010). Quantum entanglements and hauntological relations of inheritance: Dis/continuities, spacetime enfoldings, and justice-to-come. *Derrida Today, 3*(2), 240–268.

Barad, K. (2011). Erasers and erasures: Pinch's unfortunate "uncertainty principle". *Social Studies of Science, 41*(3), 443–454.

Barad, K. (2012). On touching – The inhuman that therefore I am. *Differences, 23*(3), 206–223.

Barad, K. (2014). Diffracting diffraction: Cutting together-apart. *Parallax, 20*(3), 168–187. https://doi.org/10.1080/13534645.2014.927623

Barad, K. (2017). No small matter: Mushroom clouds, ecologies of nothingness, and strange topologies of spacetimemattering. In A. Lowenhaupt Tsing, N. Bubandt, E. Gan, & H. A. Swanson (Eds.), *Arts of living on a damaged planet: Ghosts and monsters of the Anthropocene* (pp. 103–120). University of Minnesota Press.

Barfod, K. S. (2018). Maintaining mastery but feeling professionally isolated: Experienced teachers' perceptions of teaching outside the classroom. *Journal of Adventure Education and Outdoor Learning, 18*(3), 201–213.

Blom, S. (2019). *Redefining time in post-truth times: Posthumanism, auto-ethnography, and parent(ing) childhoodnature in environmental education research* [Paper presentation]. 2018 Annual meeting of the American Educational Research Association. https://doi.org/10.3102/1438900

Blom, S. M. (2020). Conceptualizing parent(ing) childhoodnature through significant life experience. In A. Cutter-Mackenzie, K. Malone, & E. Barratt Hacking (Eds.), *Research handbook on childhoodnature: Assemblages of childhood and nature research* (pp. 1–26). Springer International Publishing. https://doi.org/10.1007/978-3-319-51949-4_127-1

Blom, S. M., & Crinall, S. (2020). Growing communities in a garden undone: Worldly justice, withinness and women. *Genealogy, 4*(2), 42.

ClassDojo, I. (2012). *ClassDojo*. Retrieved January 9, 2022 from https://www.classdojo.com/

Crinall, S. (2017). *Blogging art and sustenance: Artful everyday life (making) with water*. Western Sydney University.

Crinall, S., Rowbottom, E. C., Blom, X. P. M., & Blom, S. M. (2020). A place with no time: Re-conceptualising child–adult relations during "homeschooling" in the 2020 pandemic. *Knowledge Cultures, 8*(2), 65–81.

Davies, B. (2021). *Entanglement in the world's becoming and the doing of new materialist inquiry*. Taylor & Francis Group.

Dernikos, B. P., Lesko, N., McCall, S. D., & Niccolini, A. D. (2020). Feeling education. *Mapping the affective turn in education* (pp. 3–27). Routledge.

Furman, C. E. (2022). Welcoming entanglements with ghosts: Re-turning, re-membering, and facing the incalculable in teacher education. *Contemporary Issues in Early Childhood, 23*(3), 253–264.

Glackin, M. (2018). "Control must be maintained": Exploring teachers' pedagogical practice outside the classroom. *British Journal of Sociology of Education, 39*(1), 61–76.

Gregory, M. (2011). Philosophy for children and its critics: A Mendham dialogue. *Journal of Philosophy of Education, 45*(2), 199–219.

Jacobi-Vessels, J. L. (2013). Discovering nature: The benefits of teaching outside of the classroom. *Dimensions of Early Childhood, 41*(3), 4–10.

Kahn, P. H., & Hasbach, P. H. (2013). *The rediscovery of the wild.* MIT Press.

Louv, R. (2006). *Last child in the woods: Saving our children from nature-deficit disorder.* Atlantic Books.

Malone, K. (2007). The bubble-wrap generation: Children growing up in walled gardens. *Environmental Education Research, 13*(4), 513–527.

McFarland, L., & Laird, S. G. (2020). "She's only two": Parents and educators as gatekeepers of children's opportunities for nature-based risky play. *Research handbook on childhoodnature: Assemblages of childhood and nature research* (pp. 1075–1098). Springer International Publishing.

Murris, K. (2016). *The posthuman child: Educational transformation through philosophy with picturebooks.* Routledge. https://doi.org/10.4324/9781315718002

Murris, K. (2020). Posthuman child and the diffractive teacher: Decolonizing the Nature/Culture binary. In A. Cutter-Mackenzie, K. Malone, & E. Barratt Hacking (Eds.), *Research handbook on childhoodnature: Assemblages of childhood and nature research* (pp. 1–25). Springer International Publishing. https://doi.org/10.1007/978-3-319-51949-4_7-2

Murris, K., & Borcherds, C. (2019). Childing: A different sense of time. In B. D. Hodgins (Ed.), *Feminist research for 21st-century childhoods: Common worlds methods* (pp. 197–208). Bloomsbury Publishing.

Murris, K., & Kohan, W. (2021). Troubling troubled school time: Posthuman multiple temporalities. *International Journal of Qualitative Studies in Education, 34*(7), 581–597.

Payne, P. (1997). Embodiment and environmental education. *Environmental Education Research, 3*(2), 133–153. https://doi.org/10.1080/1350462970030203

Reinhart, G. R. (1980). Social organization of ants and humans. *Free Inquiry in Creative Sociology, 8*(1), 100–103.

Rooney, T. (2016). Putting time aside: Navigating the flow of becoming in a posthuman world. *Global Studies of Childhood, 6*(2), 190–198.

Tammi, T., Rautio, P., Leinonen, R.-M., & Hohti, R. (2020). Unearthing withling(s): Children, tweezers, and worms and the emergence of joy and suffering in a kindergarten yard. *Research handbook on childhoodnature: Assemblages of childhood and nature research* (pp. 1309–1321). Springer International Publishing.

Waite, S. (2011). Teaching and learning outside the classroom: Personal values, alternative pedagogies and standards. *Education, 39*(1), 65–82.

Wildlife Information Rescue and Education Service [WIRES]. (2021, March). *Northern Rivers Snake information.* http://www.wiresnr.org/SnakeFacts.html

Wilks, R. (2014). *Aboriginal totems.* Central Tablelands Local Land Services.

Wolfe, M. (2021). *Affect and the making of the schoolgirl: A new materialist perspective on gender inequity in schools.* Routledge.

9

THE NINTH TURN

Diffractive data entanglements for presenting posthuman educational research – part two

Introduction

As discussed so far, this book explores early school years teachers' perceptions of nature through a posthuman theoretical framework. In this Turn, we continue Caitlyn's story as a diffractive exploration of a second single-lesson participation, video-recalled conversation, visual journal entries, and photographs presented as a series of five diffractive data entanglements. These follow on from the Eighth Turn where the first six diffractive data entanglements were presented.

As contextualised in the Eighth Turn, Caitlyn is a 33-year-old woman in her fifth year of teaching who identifies as a Yamatji woman of the Wajarri Nation. Caitlyn's school is located in the Northern Rivers of New South Wales and had a population of approximately 34 students in 2020. There is a teaching principal and three classes in total, and Caitlyn is teaching the combined kindergarten, grade 1, and grade 2 class.

This Turn presents the second exploration of the diffractive data entanglements (the data diffracted through the diffraction

DOI: 10.4324/9781032703473-9

gratings of childhoodnature, affective atmospheres and material-discursive practices [see the Fourth Turn]) with Caitlyn. It is an entangled and creative journey where images form part of the data and where video-stimulated recall conversation data and certain direct quotes on theory are presented as prose (inspired by the works of Crinall, 2017) to give the data a sense of flow in its becoming. The work is presented in present-tense and active voice to bring you, the reader, into the room with me. To facilitate the flow, references to Caitlyn's quotes are purposefully added as footnotes instead of in-text. These literary acts highlight that the words on the page are "not inert, passive objects in the service, beautification and edification of humanity [in the relationship with the non-human world] … but the lively 'stuff' of matter that is articulate and agential" (Davies, 2021, p. 57). I invite you, the reader, to undertake your own intra-active journey of data entanglements with these data as you read.

Enacting diffractive data entanglements follows the first three steps proposed in the Sixth Turn:

1. On completion of phases 1 and 2 of the data entanglement process (see the Fifth Turn), I re-experience the documented data through a deep dive into watching the video recording of the lesson participation (data from phase 1), looking at the photos from the lesson participation (data from phase 1), listening to the audio-recordings of video-stimulated recall conversation (data from phase 2), and entangling the teachers and my own visual-journal (data from phase 3). While re-engaging with data, I attune to the concepts of childhoodnature, affective atmospheres and material-discursive practices to explore how these concepts grow and expand the data in "new" and unexpected ways.

2. While re-engaging in the data, further field notes and visual-journal entries document how the data's agency emerges – how the data speaks and makes itself known and in doing so, how further data is generated.

3. The data with each participant is presented individually through patterning from the theoretical diffraction gratings (childhood-nature, affective atmospheres and material-discursive practices).

From this process, eleven diffractive data entanglements became known. They are presented across two Turns – differentiated by different locations in time, and therefore, space (Barad, 2007). In this Turn, we present the second five diffractive data entanglements following on from the five presented in the Eighth Turn. These diffractive data entanglements make lively the ReTurning Learning theoretical framework by giving voice to childhoodnature, affective atmospheres and material-discursive practices through the enactments of Caitlyn, a classroom teacher, in an early school years classroom in our two lesson participations and video-recalled conversations.

Diffractive data entanglement #7: the flow of student-centred response-ability

Later, in our video-stimulated recall conversation, Caitlyn shares about authentic student-led pedagogies and our response-ability as teachers:

We were doing an art lesson,
we had done math and we all went outside just to do our art lesson outside,
and we were drawing landscapes and things like that
and then they tied in the math,
"oh these are all 2D shapes that we're doing in our artwork"

FIGURE 9.1 Caitlyn's visual-journal entry: symbolising growth; personally through the palm centre and collectively through the Aboriginal symbols sitting around the outside of the palm watching growth happen through the centre; representing community and people. *Original artwork by research participant "Caitlyn". Reproduced with permission.*

So then we scrapped the rest of the math lesson
and then went out and did art because that's what the kids –
but they still learnt it,
but it went more in their direction,
they controlled it a little bit more.
And I think,
well, I'm not really teaching them am I [Chuckles],
I'm just guiding them through the process and providing them
* the space*
and the tools to do it.
I think that's when I feel better as a teacher;
I'm like yeah that was awesome…
it's high engagement,
and I call them little lightbulb moments,
you know you see them just going off all the time,
and then they start to make different connections
and they talk to each other about different connections
and they get it more.[1]

Perhaps this is the response-ability of each teacher to demonstrate their ability to enact this kind of responsive pedagogy with our students and therefore, to child body *as nature*. Childhoodnature here is an expression of flow, an affective response, and an acknowledgement of the cycles of nature that includes all (more than human) bodies.

I reflect on a conversation I had with colleagues, thinking with Barad (2007), responsibility environmental education. My colleague Shae (Brown et al., 2020, *p. 222; emphasis in original*) shares,

> *I understand agential realism as*
> *"an appreciation of the intertwining of*
> *ethics, knowing, and being". (Barad, 2007, p. 185)*

It is an inclusive inseparability
that Barad describes as
an ethico-eco-onto-epistemology.
*This intertwining brings **responsibility,***
so there is always an ethical dimension
to everything
we do in the world.
Barad describes all knowledge making,
and all actions really,
as having a material effect,
because we are entangled;
inseparable with/in a co-generative world.
Life is not all happening "out there",
happening is ubiquitous everywhere,
all the time.
In this way all elements of life,
across the boundaries of human and nonhuman,
living and so-called non-living,
all of it,
have agency in a way that is active and entangled.
It is so different from
the perspective of classical science
where the world is "made" of
interacting but
still separate parts.

In thinking with responsibility, I diffract from the "out there" in this passage that "glows" as an affective, embodied sense (MacLure, 2013). From these words on the page from Shae, I return to the work from the Eighth Turn and think about the "out there" when I thought with community with Sarah,

> "out-there" to also be "in-here";
> where the "in-here"
> becomes what "matters"
> Blom (2020); Kahn & Hasbach (2013); Payne (1997)…

> This idea of community
> does not need physical bodies
> to come together,
> enabling the growing together,
> but seeks to explore the potential
> of (be)coming together
> with a withinness.
> Blom & Crinall (2020, p. 15)

Perhaps the withinness is now a "within-us" as we each hold our response-ability for our human body *as nature*. I recently heard Wiradjuri woman, Brenda Matthews, speak about this responsibility – to dive deep within and heal there first, as she experienced being stolen from her family but it was not officially classified as such (Matthews, 2024). Perhaps Cassie has done this work and is now acting – and activating – this response-ability to enact from within as an intra-action to empower and ignite the small bodies of children to response-ability too.

Diffractive data entanglement #8: affective atmospheres and the micro

It is June now, 2020, in week 3 of term 2. This morning, I arrive for the morning literacy lesson. The weather has turned colder, however, in Northern NSW, this is still relatively warm with an overnight low of 17°C, a top of 23°C, and sunny. The most significant change is that since the last visit, the COVID pandemic has started. NSW schools have gone into lockdown at the end of term 1 for about four weeks and in term 2, schools have returned to mostly business as usual with the added precautions of hand sanitiser and more restricted access for guests. Although the impact of COVID on this research, and on everyone in the context of this work runs much deeper than described here, I return the focus to the micro-world of Caitlyn's classroom in acknowledgement of the greater worldly happenings that were certainly felt but did not make themselves known through this diffractive data entanglement.

In Caitlyn's classroom, while most of the students are working in groups completing their work, Caitlyn takes turns working one-on-one with each student to gather an assessment about their reading skills. Caitlyn is sitting on an exercise ball, which serves as the teacher's chair, and she and the student are sitting opposite each other at a kidney-bean table. The affective atmosphere of the classroom has a quiet, gentle hum of students going about their work; I consider this as the "productive work hum" that teachers often strive for. The affective atmosphere is calm, and the students do not seem to want to disrupt that, except for the occasional sound of the electric sharpener. As Caitlyn finishes with the second child, they reluctantly leave the kidney-bean table to return to their desk to work on their writing task. The next student is already approaching the table, ready to sit down by the time the child before stands up to move. Even though Caitlyn is very much present and with the child at the kidney-bean desk and listening to their reading, she is equally across what the other students are doing around the room. She vocally alerts them to her aware-ness, and they immediately get back on task. I ponder on the role of the affective atmosphere of the classroom and Caitlyn's ability to perceive these disturbances – the children being distracted or off-task – through her capacity to feel and sense the room rather than by just *seeing* what the children are doing. This attunement to the affective atmosphere of the classroom extends beyond the human-centric ideas of the classroom or school climate to include the material-discursive forces of the nonhuman actants and imma-terial forces that are active contributors (Verlie & Blom, 2021).

Caitlyn made a different observation about the affective atmos-phere of the room through the lens of one of her students. Caitlyn shares with me that one of the children is really affected by the research cameras in the room. Caitlyn has done some filming in the past and describes this student as *"a deer in headlights"* due to anxiety. Caitlyn describes this anxiety as the "atmosphere" for this student. Caitlyn suggests that it is because the student is *"really, really afraid of getting things wrong"*. Caitlyn has a class-room saying that "mistakes are gold" so students are taught to leave their written mistakes in their workbooks and that there is "no thing that is wrong". Caitlyn has seen the progress for this student but cannot speculate why they would feel this way in their first year of schooling. Caitlyn acknowledges that the anxiety has decreased during the year but is exacerbated by dif-ferences in the classroom setting such as having the research cameras in the room. I take note of how subjective the affective atmospheres of the room are for each participant.

The fourth child comes and sits at the kidney-bean table. Some of the other children have finished their sentences and are moving on to the next task. One of the children approaches

Caitlyn with a question from their next task and Caitlyn directs them to another child in the classroom. Caitlyn has established a classroom culture where peer learning is valued and normalised. The children are not perturbed by being re-directed nor by being asked to act as the "more-knowledgeable-other".

The respect the students hold for Caitlyn is as evident as the respect she holds for them. This is shown through seemingly small actions such as never writing in students' workbooks when she marks them because she honours that the books belong to the students. Instead, Caitlyn makes sticky notes to add to their work with feedback. Caitlyn shares that as teachers,

> *we don't have the right*
> *to go scribbling over their hard work*
> *and all their thinking,*
> *so put it on a sticky note*
> *so when they're done with it*
> *they can rip it out.*[2]

Looking at the learning scape from this position as the researcher reviewing the video of the lesson participation, through the lens of the GoPro camera, the artificial glow of the fluoro lights speaks loudly. Their presence in the room, while barely noticeable at the time, now stands out like a neon sign or a bright full moon at night. I ponder how much these lights are "rearranging the furniture" and other bodies through the agential forces that they operate at micro scales (see Barad, 2007). I return to the passage from Barad (2007, p. 108),

> When light bounces off
> a relatively large object,

the disturbance it imparts
is negligible
relative to the accuracy of the measurement.
That is,
it is often the case
that any such disturbance
is too small to notice.
(For example,
we don't notice the furniture
being rearranged in the room
when we turn a light on in a dark room,
although this is strictly the case.).

These business-as-usual approaches to life, which we rarely if ever stop to consider, are not invisible nor silent nor make-believe. They are real, they have agency, and they make an impact. Perhaps even more so through the sensitive bodies of children. The impacts of a fluoro light in the classroom that does not emanate the same quality as the sun. However, we need and appreciate the fluoro light for the purpose it serves. However, how often do we talk about these material affects with each other? How often do we turn to children to understand what they are sensing, what they are feeling and how they can *be-with* those sensitivities in life? These conversations with children are rich, unfiltered, and insightful.

Diffractive data entanglement #9: paper–plastic hybrid, indeterminacy, unnoticed

A child gets up to go and do something and as they do, a piece of paper wrapped in a plastic pocket flies from their desk and lands on the floor. They do not seem to notice; no one else does either. I notice the other scattered pieces of paper around the room (see Figure 9.2).

FIGURE 9.2 The paper on the classroom floor. *Photographs taken and arranged by Simone Blom.*

Despite all the movement by the children around the room, no one does anything with the pieces of paper, however, the children do ensure that they do not step on them, so there is an awareness that they are there. I wonder how they got there, who is and should be responsible for them, and how these papers on the floor are akin to the rubbish in the schoolyard. I consider why – when thinking of children and messy schoolyards – do they think it is someone else's responsibility to clean up the mess?

Caitlyn finishes with the child at the kidney-bean table, calls for everyone to bring their work onto the floor and outlines what they will be doing for the rest of the lesson. Once seated, Caitlyn calls upon two students to go around and pick up everything that is on the floor. The students do so without question. However, they leave one. They pick up the pieces near their table but do not touch the piece of paper wrapped in plastic.

I consider what further approaches would support children in taking response-ability with their environment, no matter where they are. There are so many small practices to build children's capacity to encourage children to take agency and responsibility for themselves and their things, including own-ership of nonhuman nature. What if children knew that they are responsible for nonhuman nature too – just like their own bodies-as-nature? Would such a practice enact space for child-hoodnature to flourish?

Caitlyn starts to introduce the topic of the "Australian Aboriginal Flag". The children are told that they are to choose what they con-sider the most important three facts are and rewrite them into an information text. In our video-stimulated recall conversation, we discuss facts.

Caitlyn says,

What is a fact, who tells us that that's a fact?

She then grapples with a student who has yet to return to the school after the COVID lockdowns. She understands the parents' worry and concern but is also concerned about the student's learn-ing as they've had ten unexplained absences so far in the year.

Caitlyn confides:

But I get her position
because she's really worried,
but there's no room for your own personal worry
when there's a Department
and a government telling you what to do…
I was told by the Department well
you're not allowed to take your leave,
you're not allowed to take any family leave;
you have to take leave without pay from your position
if you're going to be caring for children,
because you're an essential worker
you have to send your kids to school full-time.
And it so was out of alignment with just who I am
as a parent and my free choice,
I thought well who are you to tell me that those kids aren't
    allowed at school
but my children are allowed at school,
it just was a really interesting period of time
and it just caused heaps of unrest.
And I did,
I just went okay I'll take leave without pay,
because I feel strongly that I shouldn't have to put my children
    in that position
so I'm not going to,
which then puts us at a loss.
So I understand that parent's point of view because she feels so
    strongly about it.[3]

Karin Murris (2020) reveals that the "pandemic is a reminder that the world is indeterminate" (p. 17). Karen Barad frequently refers to indeterminacy. Indeed, it is central to much of what we know and therefore, don't know, about everything. Barad (2018) presents,

The indeterminacy of
space, time, and matter
at the core of quantum field theory
troubles the scalar distinction
between the world of subatomic particles
and that of colonialism,
war,
nuclear physics research,
and
environmental destruction.
p. 64

This "lively indeterminacy lives in, around and through us – that will help us face the depths of what responsibility entails" (p. 81). The intra-connections of the micro and macro worlds give a deeper understanding about one another. Considering environmental destruction, responsibility, COVID and the paper-plastic hybrid not as separate entities but as entanglements that have their own story to share, that contribute to how we, as humans, understand each other and the world. This is our collective response-ability to enact and move our human bodies in a way that is respectful and honours the body-as-nature that communicates so clearly and directly with us, constantly. "How might we approach the possibility of listening?" (Barad, 2018, p. 77).

FIGURE 9.3 The paper-plastic hybrid remains. *Photograph by Simone Blom.*

The children move to their desks to do their work and the plastic wrapped paper remains on the floor (Figure 9.3).

Caitlyn comments to no one in particular, *"it's getting toasty in here"* and grabs the remote control for the split system to turn down the temperature. As the children are getting organised, she provides praise to one of the students, *"[student name] has already started writing, really on task [student name], well done".* The impact this has on the classroom atmosphere is evident. The other children prick up their ears and look around to see the student at work while becoming a little more focused and purposeful to get started. Caitlyn attends to a group of students who are slow to get started. She stays with them and encourages them to continue with their work. Caitlyn moves to speak to the other group of students working on the other side of the room. On the way, she picks up the paper-plastic hybrid (see Figure 9.4). I wonder if adults create a culture where children feel that they do not need to look after anything because the adults are always there to pick up the pieces? How can an adult's response-ability can be reconceptualised? I wonder if children were taught to look after and respect their internal environments such as their bodies, bedrooms and

FIGURE 9.4   Picking up the pieces. *Photograph by Simone Blom.*

classrooms from an affective perspective, if this care and respect would carry out into the natural environment? I ponder if it's possible for the quantum mechanics principle of the micro affecting the macro to be applied to this situation and the impacts that would have on our collective relationship with nonhuman/human nature.

Diffractive data entanglement #10: the teacher's chair

Caitlyn moves around the room checking on the other group of students. She returns to the original table and asks the child with a plaster cast how they are feeling. Caitlyn offers to type up their writing. She asks if he is frustrated, and he nods. Caitlyn directs all the children to the floor in preparation for recess time. In their exuberance, they do not push in their chairs. With a gentle reminder from Caitlyn, they race back to their tables to push in their chairs; Caitlyn reminds them to walk not run. One chair remains out from the table; Caitlyn was the last one sitting in it. One of the children does not just push in their own chair and race back to the floor, but bounces back and goes around the room, arriving back at the table to push in Caitlyn's chair. It is the child with the plaster cast. I wonder if it is the moment of connection they share with Caitlyn that leads the child to move in this way and to have the want

FIGURE 9.5 A photo story of Caitlyn's chair. *Photographs taken and arranged by Simone Blom.*

and responsibility to push in Caitlyn's chair (see Figure 9.5). The human-human relational action speaks more loudly than words.

Taylor (2013) writes diffractively about the "teacher's chair". However, in Caitlyn's classroom, it is a different kind of teacher's chair. It is not different to the children's chairs, and it does not sit majestically out the front of the classroom as a "magisterial chair". It becomes the teacher's chair incidentally through a movement. But it emanates – it has a material-discursive force – that perhaps propels a child's body to swiftly change their movement and circle back to push in the teacher's chair. I wonder if this response-ability to the order of the classroom is inspired by relationality. I consider if the point of connection – the intra-activity between teacher-child bodies where time is nothing - enacts a change in movements for the child.

Taylor (2013) observes how "chair, objects and bodies are entangled intra-active forces" (p. 694) as an example of the point made by Barad (2007) that "bodies in the making are never separate from their apparatuses of bodily production" (p. 159). Unlike "Malky's" chair which Taylor (2013) describes, Caitlyn's chair is a fluid part of the classroom space. It is not owned by Caitlyn – it is just a chair she sits in during that particular lesson. Caitlyn's chair is designed for small bodies and is not differentiated and does not stand out in the classroom space. It is not "elegant, black and chrome....on casters and with a tilting facility..." (Taylor, 2013, p. 613). On the contrary, Caitlyn's chair could not be more mundane, more the same as any other chair in the room, the school and most likely, in at least all Government schools in the state. So, what makes this chair different? What material-discursive force from the chair calls the student to come and push it in? Although the façade of the chair does not look different from the outside, there is some-thing about the chair that makes it stand out and that is different to that child. The enactment of pushing in the chair is not fuelled by promises of rewards such as stickers, Class Dojo points, or verbal praise, but appears in response to the connection with his teacher. This point of connection – the enactment of relationality – dem-onstrates the reconfiguring of the particles at the micro-level (see Barad, 2007, p. 150).

All the children gather on the floor and Caitlyn asks three children to share. After their discussion, one of the children announces, "it's recess!" The children are eager to run out-side but wait respectfully for Caitlyn's instructions. She adds, *"have a look around the room and make sure everything is back where it started. Make sure your belongings are put away. Make sure your table is neat and tidy. Make sure I have your writing. Thank you."* The children move from the carpet to make some adjustments to the classroom. However, the child with the plas-ter cast stays on the floor with Caitlyn and shares some con-cerns about going out into the playground. Caitlyn shares that it is important for the child to get some fresh air and says that the child does not have to play; that she could give them some colouring in to do quietly, *"Just see how you feel okay and let me know if you're starting to feel worse"*. The lesson completes.

Diffractive data entanglement #11: sustainability, nature, rainbows

When reflecting on sustainability in our video-stimulated recall conversation, Caitlyn shares,

> *Science at the moment*
> *is all based around sustainability and growth,*
> *so I kind of think yeah I'm addressing it*
> *but do I know that it's sustainability?*
> *It still feels detached from the whole concept*
> *even though it's all about conservation and sustainability.*
> *I don't know,*
> *there still feels like there's some mismatch…*
> *and maybe on reflection after this unit*
> *I'll probably feel like oh we did that actually really well*
> *but the intention to teach it in the beginning doesn't feel strong,*
> *there's got to be a different way*
> *that's more meaningful and not just this same-*
> *it just doesn't feel very ingenious.*
> *So maybe that's where the mismatch is,*
> *is that you're trying to reinvent the wheel*
> *to think differently about it [how to fit sustainability in the*
>   *curriculum]…*
> *I don't want to bag out the curriculum*
> *and the expected outcomes and all of that,*
> *but just making it more meaningful in the classroom…*
> *there's no groundwork for how sustainability fits in there.*[4]

Caitlyn then moves to discuss her perceptions of nature, which she contextualises for the situation with COVID (see Figure 9.6),

I guess probably it'd be good to talk in a covid context with that,
it was so important for well-being
that we just encouraged everyone to be in nature.
And I just thought isn't that interesting,
we're sending all this work home to our kids
but we're really emphasising get outdoors,
go for a walk get some fresh air,
get outside and be in nature,
because that's so important at this time,
so why don't we do that all the time.
And in my journal, I wrote a reflection…
because you came and did the observation,
and then I picked my son up from school that afternoon,
I know that nature is vital to his well-being…
He's an outdoors kid.
If he can just be outside
playing with the chickens
and fishing
and catching lizards
and looking after frogs all day,
he's just the happiest human being in the world.
Whereas he's in trouble at school all the time,
and then they put him on a behaviour plan
and took him out of the playground,
and I just thought what are you doing –
but you can't argue like
you're taking the nature out of my child
but they are.[5]

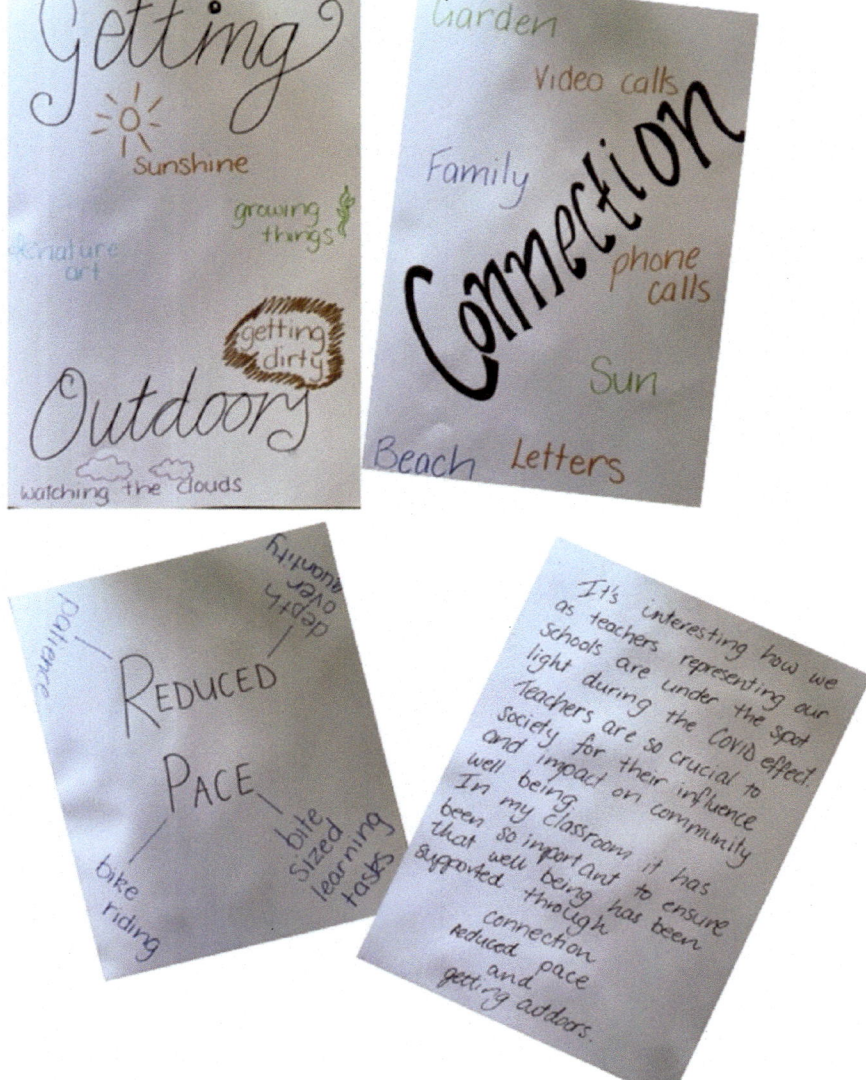

FIGURE 9.6 Caitlyn's visual-journal entries: A collage of pages reflecting on nature. *Original artwork by research participant "Caitlyn". Reproduced with permission.*

Being outdoors exploring nature
allows him to connect with himself.
He has settled into Year 1 really well
but he has a stronger urge for technology,
gaming, YouTube and devices.
I can't help but wonder
has school begun the process of
conforming his connection to nature out of him.[6]
And I thought
how are you going to expect to get results
with a kid who needs to be outdoors,
and your punishment,
your consequence
is punishing him
by taking him out of the playground
and locking him inside.
So for me my reflection
on how important nature is after that was like
how important it is to my kids,
and how it is taken away,
it's almost a reward to be outdoors,
it's a reward to go out
and play,
it's a reward to go to the beach,
it shouldn't be... [see Figure 9.7]

I had an awful childhood really to be honest,
like lots of domestic violence and abuse
and all sorts of stuff,
so I guess being outdoors was kind of a release.

FIGURE 9.7 Caitlyn's visual-journal entry: reflecting on teaching, nature and growth. *Original artwork by research participant "Caitlyn". Reproduced with permission.*

Oh we used to go disappear,
our house backed onto a creek
and that was my most memorable times,
just disappearing with my siblings.
And we'd be gone all day
from one end of the creek all the way,
through the back of all these houses up to the other end,
we'd find eels and turtles,
like that was reprieve
from the rest of life,
just to disappear,
where we could just go play
and be kids
and explore…
I just find it [nature] humbling for me…
I try to find some connection as well
because my family is from the
Yamatji tribe in Western Australia…
it's kind of coastal up there as well,
and I think maybe there's something ingrained in me
that just has this draw to it [the ocean].
Like the Willie wagtails are our totem
and that's crazy important to my boys and I,
if we see a Willie wagtail
we really watch the Willie wagtail,
we listen to it
and pay attention to what it's doing
and wonder how that might impact our day.
which sounds ridiculous to other people,
but it's really important to us.

And seeing the wonder in it all,
like the rainbows [Chuckles],
and a lot of people,
I don't know
maybe they're disconnected,
maybe I'm not crazy,
there's disconnection out there…
My son he's 10
but if he sees a rainbow
he takes a photo
and he sends it to me every time
because he's like "there's a rainbow mum!"
and I'm like awesome…
that's so cool
that a 10-year old is stopping
and noticing the rainbows.[7]

Conclusion

This Turn provides further evidence of the ReTurning Learning theoretical framework in practice, by detailing the classroom happenings of one teacher, Caitlyn, and her students in an everyday lesson. Through the presentation of a series of diffractive data entanglements of lesson participations, video-stimulated recall conversations, visual-journal data and emails, I thought with the research question to explore Caitlyn's perception of nature and how this informs her pedagogy.

Caitlyn perceives nature as an entanglement of her relationship with herself, nonhuman nature and culture. As a Yamatji woman, Caitlyn sees nonhuman nature as a place of connection, with self, each-other and the nonhuman. Caitlyn has an ancient connection with Country through her ancestors and values and has a deep respect for nonhuman nature.

Caitlyn puts childhoodnature to work in authentic ways of teaching children how to have conversations together – rather than relying on the teacher to guide discussions, joining the students on the floor for class time together, incorporating genuine experiences *in* nature into the flow of her everyday lessons, and experimenting with a different kind of lesson planning and pedagogical practice: one that responds to the flow of the students' ideas rather than the material-discursive force of curriculum and syllabus outcomes. Caitlyn attunes to the affective atmosphere of her classroom to respectfully enact her teacher response-ability by making movements where they are called for. In doing so, childhoodnature practices are also enacted, as child bodies are sensorily responded to.

Caitlyn's contribution is that being aware of the differences between adult and child bodies in a way that disrupts traditional classroom practices of separating teachers and students, changes the relational focus from "top-down" to "next to". Environmental education has a strong relational focus that is deeply entangled with First Nations ways of knowing and being where education *about, in, with* and, at times, *as* nature is practised.

Notes

1 Caitlyn, video-stimulated recall conversation, 19th February 2020.
2 Caitlyn, video-stimulated recall conversation, 19th February 2020.
3 Caitlyn, video-stimulated recall conversation, 11th June 2020.
4 Caitlyn, video-stimulated recall conversation, 11th June 2020.
5 Caitlyn, video-stimulated recall conversation, 11th June 2020.
6 Caitlyn, visual-journal entry.
7 Caitlyn, video-stimulated recall conversation, 11th June 2020.

References

Barad, K. (2007). *Meeting the universe halfway: Quantum physics and the entanglement of matter and meaning.* Duke University Press.

Barad, K. (2018). Troubling time/s and ecologies of nothingness: Re-turning, re-membering, and facing the incalculable. In M. Fritsch, P. Lynes, D. Wood, K. Barad, T. Clark, & C. Colebrook (Eds.), *Eco-deconstruction: Derrida and environmental philosophy.* Fordham University Press.

Blom, S. M. (2020). Conceptualizing parent(ing) childhoodnature through significant life experience. In A. Cutter-Mackenzie, K. Malone, & E. Barratt Hacking (Eds.), *Research handbook on childhoodnature: Assemblages of childhood and nature research* (pp. 1–26). Springer International Publishing. https://doi.org/10.1007/978-3-319-51949-4_127-1

Blom, S. M., & Crinall, S. (2020). Growing communities in a garden undone: Worldly justice, withinness and women. *Genealogy, 4*(2), 42.

Brown, S. L., Siegel, L., & Blom, S. M. (2020). Entanglements of matter and meaning: The importance of the philosophy of Karen Barad for environmental education. *Australian Journal of Environmental Education, 36*(3), 219–233.

Crinall, S. (2017). *Blogging art and sustenance: Artful everyday life (making) with water.* Western Sydney University.

Davies, B. (2021). *Entanglement in the world's becoming and the doing of new materialist inquiry.* Taylor & Francis Group.

Kahn, P. H., & Hasbach, P. H. (2013). *The rediscovery of the wild.* MIT Press.

MacLure, M. (2013). Researching without representation? Language and materiality in post-qualitative methodology. *International Journal of Qualitative Studies in Education, 26*(6), 658–667. https://doi.org/10.1080/09518398.2013.788755

Matthews, B. (2024). *Brenda Matthews.* Dean Buchanan, DRB Entertainment. https://www.brendamatthews.com.au/

Murris, K. (2020). Introduction: Making kin: Postqualitative, new materialist and critical posthumanist research. In *Navigating the postqualitative, new materialist and critical posthumanist terrain across disciplines* (pp. 1–21). Routledge.

Payne, P. (1997). Embodiment and environmental education. *Environmental Education Research*, 3(2), 133–153. https://doi.org/10.1080/1350462970030203

Taylor, C. A. (2013). Objects, bodies and space: Gender and embodied practices of mattering in the classroom. *Gender and Education*, 25(6), 688–703. https://doi.org/10.1080/09540253.2013.834864

Verlie, B., & Blom, S. M. (2021). Education in a changing climate: Reconceptualising school and classroom climate through the fiery atmos-fears of Australia's Black summer. *Children's Geographies*, 1–15. https://doi.org/10.1080/14733285.2021.1948504

10

THE TENTH TURN

Meaning-making with/as nature for educational futures

Introduction

Throughout this book, we have explored early school years teachers' perceptions of nature and how this informed their pedagogy through the posthuman concepts of material-discursive practices, affective atmospheres and childhoodnature. To enact diffractive data entanglements, I introduced both diffractive ethnography to the field of environmental education research and transqualitative inquiry to research more broadly. More specifically, we have focused on the question,

What are early school years teachers' perceptions of nature and how do they inform pedagogy?

In the Eighth and Ninth Turns, the data was presented as diffractive data entanglements with teacher participant, Caitlyn. The diffractive data entanglements enabled deep engagement and consideration as part of the ethnographic tradition and allowed space for the authentic practice of listening to teachers. Listening

DOI: 10.4324/9781032703473-10

to teachers was identified as lacking in much of the research, identifying that generally research is "done *to* teachers, rather than *with* them" (Lingard et al., 2003, p. 403; italics in original). Moreover, by having a single Turn dedicated to Caitlyn's teacher data, the nonhuman actants are also given space to be included, illuminating the richness of more than "one person's story", while acknowledging the agency of the entangled materiality of human and nonhuman bodies that constitute each classroom.

In this Turn, the diffractive data entanglements are thought with the extant research literature through deep engagement and further entanglement to contextualise the data within existing environmental education research.

Learnings with Caitlyn: returning to the extant literature
Entangled perceptions with/as nature

Past research studies have identified a range of categories to thematically analyse teachers' perceptions of nature (Bonnett, 2002; Cutter-Mackenzie & Smith, 2003; Desjean-Perrotta et al., 2008; Ernst, 2014; Fraser et al., 2015; Kimaryo, 2011; Thi To Khuyen et al., 2020; Torquati et al., 2013), where "perceptions" included conceptualisations, perspectives, attitudes, beliefs, difficulties and self-efficacy, and "of nature" included "of the environment", "of sustainability" but rarely "with" and never "as". Teachers included preservice and in-service teachers. Some of these concepts are now explored through and with Caitlyn's data.

Posthuman leanings through relationality

Cutter-Mackenzie and Smith (2003) asserted that most teachers in their study held eco-socialist (communalism) perspectives which were described as accommodation and "valuing life above all". This eco-socialist view was demonstrated at times by Caitlyn, while at other times, the data leaned into the posthuman. Fraser et al. (2015) identified a kind of "spiritual thinking" in environmental education research which they described as surprising but not uncommon. This perception of nature was evident in Caitlyn's data as she discusses her connection to Country through listening to the nature of the Willie Wagtail,

> ...*if we see a Willie Wagtail*
> *we really watch the Willie Wagtail,*
> *we listen to it*
> *and pay attention to what it's doing*
> *and wonder how that might impact our day.*
> *which sounds ridiculous to other people,*
> *but it's really important to us.*
> *And seeing the wonder in it all...*

More recent posthuman research has described nature as the "diffractive teacher" (Murris, 2020), as the teacher (Blom, 2016) and "co-teacher" (Blenkinsop et al., 2020). Blenkinsop et al. (2020) described a process of inviting nonhuman nature to be co-teacher, which involved careful and attentive listening to the environment and in doing so, nature's agency is enacted and it "can become an active member in teaching and learning" (Blenkinsop et al., 2020, p. 459). Despite these posthuman leanings, which extend beyond the traditional human-nature dualism, the focus is (still) human-centred: created for human consumption where humans are central to the product.

In a similar relational paradigm, Carson (1965) instructed parents to "take time to listen and talk about the voices of the earth and what they mean" (p. 68), Blenkinsop and Piersol (2013) asserted that research involving listening to the natural world has generally been confined to spiritual and First Nations ways of knowing. Relationality was also demonstrated through the importance of time *in* and *with* nature; something external (Loughland et al., 2002). The view of nature being "out-there" was first described by Payne (1998) and has been echoed by many environmental education researchers since (Blom, 2020; Loughland et al., 2002; Shepardson, 2005). The externalisation of the natural world is one of the dominant discourses and perceptions held by adults, as has been shown through the data (Rousell & Cutter-Mackenzie-Knowles, 2020). Despite the posthuman leanings of Caitlyn's perceptions of nature, the clear relational foci (Loughland et al., 2002) are more aligned with traditional conceptualisations of humans in relation *to* nature and therefore, are (still) considered anthropocentric or technocentric ('Cutter-Mackenzie & Smith, 2003; O'Riordan, 1977; Quinn et al., 2016).

Torquati et al. (2013) described one way of conceptualising nature that they observed in their study as "we are part of nature; nature is all around us; nature encompasses all" (p. 733); this both promotes and negates the posthuman concept of human *as* nature. This category was demonstrated through Caitlyn's evident lean into a posthuman perception of human *as* nature through her verbal conceptualisations of nature shared during the video-stimulated recall conversations, highlighting the idea that "we are part of nature" (Torquati et al., 2013, p. 733). However, as demonstrated by Torquati et al. (2013) where participants had identified humans both "as" nature and separate from it in the one sentence, there were similar instances with Caitlyn where the data emphasised the perception of *in* nature, *"we're really emphasising get outdoors, go for a walk get some fresh air, get outside and be in nature"* (Caitlyn, 11th June 2020).

A relationship of deep appreciation

In contrast to the "relational" focus, was the "object" focus as proposed by Loughland et al. (2002), which described the environment as a place that contained living things. Although Caitlyn discussed objects *in* nature, there was always a foregrounding of the relational aspect *with* nature. Through Caitlyn's vignette on bees, wasps and dragonflies, the culture of respect, awe, wonder, and magic she has embodied is evident. The purpose of this research was not to identify the source of this relationship, but she did hint at the cultural connection to dragonflies in her vignette. The appreciation and as such, the awe, wonder and magic of nature were reminiscent of the work of Carson (1965) who proposed that "if a child is to keep alive his [sic] inborn sense of wonder … he [sic] needs the companionship of at least one adult who can share it, rediscovering with him [sic] the joy, excitement and mystery of the world we live in" (p. 45) in encouraging parents to get outside with their children (see also Greenwood, 2020).

Weston (2004) stated that in education, teachers need "more than ourselves to 'go wild' … we need the presence of the more-than-human world" (p. 39). In favour of this view, Sobel (1996) claimed that adults need to give children "time to connect with nature and love the Earth before we ask them to *save it*" (p. 10; emphasis added), and Kalvaitis and Monhardt (2015) further asserted that effect is required to motivate people to act in environmental education. Caitlyn modelled these philosophies and research findings in her lessons with her students. Caitlyn thrived in sharing her deep adoration for nature through creative and innovative learning experiences and lessons. While the three paradigms of *in, for* and *about* were evident in Caitlyn's pedagogy, time *in* nature was the dominant practice. Time *in* wild nature or education *in* the environment is still purported by the research as providing children with the best access to become environmental stewards and eco-literate citizens (Blenkinsop et al., 2020; Malone, 2007; Sobel, 2008; Weston, 2004), although even the concept of "humans *in* nature" adopts anthropocentric (technocentric) thinking and theorising, and promotes childhood, and children, and nature as binary.

The movement to learning *in* nature, which could be argued to have begun with the "new nature movement" (Louv, 2006) and getting children "outside", needs to look to the details of how teachers embody and facilitate the concept of "human bodies *as* nature". Rather than "education *in* nature" being seen as a beneficial act of restoration and connection, outdoor experiences can be reconceptualised as an enactment of returning the human body to a place where it is naturally from and belongs.

The discrepancy between perception and practice has not been widely reported in environmental education research (Simmons, 1993), with much attention focussed on investigating *why* practitioners have ecological tendencies rather than *how* these tendencies influence their pedagogy, such as through the field of significant life experience (see for example Chawla, 1998). Previous studies in Sustainability and STEM education have ascertained that teachers' practices are influenced by teachers' perceptions (García-González et al., 2020; Simmons, 1993; Thi To Khuyen et al., 2020), however, there is a deficit of environmental education research that specifically identifies how teachers' perceptions of nature inform their pedagogy. The findings from this study suggest that teachers' perceptions of nature (still) come from a human-centric position, despite leaning into some posthuman ideas. Nature is perceived as something external, "out-there" and as a resource: not in a destructive way, but as a place to appreciate and recharge. From this perception, this teacher (still) demonstrated education *about, for,* and *in* the environment with little to no explicit education *with/as* nature.

Posthuman theoretical diffractions

As highlighted at the start of this study in the First Turn, the significance of this research lies in the adoption of posthuman theorising as there is a dearth of research in environmental education about early school years teachers' perceptions of nature from a posthuman perspective. Moreover, I argued that more research is needed to understand how posthuman concepts are practised in classrooms through teachers' pedagogies. The Returning Learning theoretical framework was developed (see the Third Turn) as a specific kind of posthuman engagement for this study as an entanglement of childhoodnature, material-discursive practices and affective atmospheres (see Figure 10.1).

The Returning Learning diffractive grating
Childhoodnature, affective atmospheres, and material-discursive practices

Through this research, there were myriad ways in which Caitlyn demonstrated an understanding of childhoodnature as implied through her perception of nature. Moreover, Caitlyn intrinsically practised childhoodnature as described in this section through the concepts of material-discursive practices and affective atmospheres.

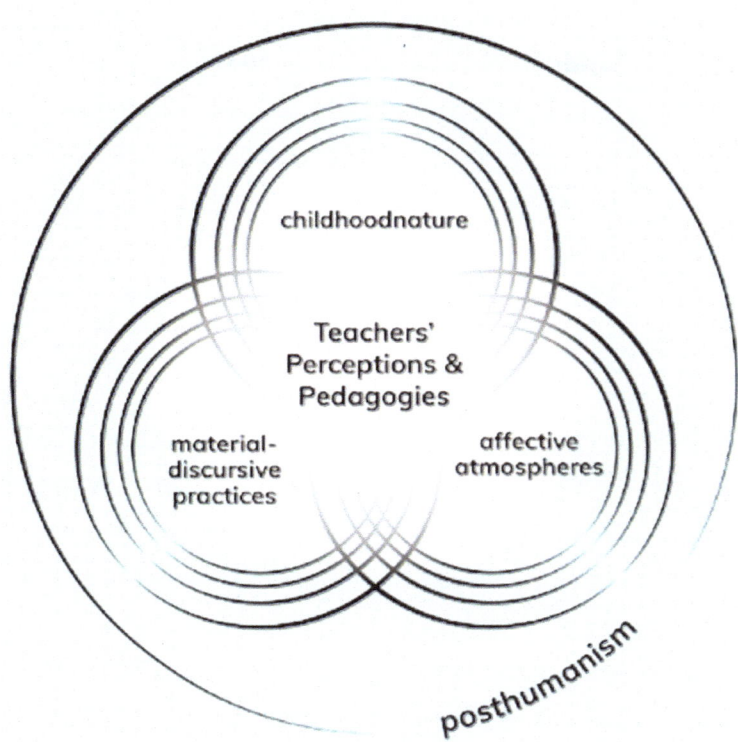

FIGURE 10.1 The Returning Learning theoretical framework. *Illustration by Kelly @ Kelly Designs for the author.*

Childhoodnature was evident in Caitlyn's practice of enacting students' agency and disrupting misopedy (Murris, 2020). These agency-enabling practices, such as sitting together with students on the floor, demonstrated the intent of fracturing the teacher/ student binary. Through troubling the teacher/student and adult/ child binary, Caitlyn enacted childhoodnature, as an act of ethical responsibility (Murris, 2020). Childhoodnature resists alignment to the material-discursive forces such as the social normativities that instruct teachers to practice acts that perpetuate the teacher/student divide. Childhoodnature challenges traditional ideas of the role and position of the teacher as a didactic instructor who stands tall, towering over the student as the all-knowing other. Moreover, these movements push back against the material-discursive practice that says "teachers sit in the teacher chair" or the "thing power" of the teacher chair (Bennett, 2010; Taylor, 2013). These pedagogical acts are gestured movements that break down the physical divide of adult and child bodies and demonstrate childhooodnature to address inequalities. In doing so, these teachers are performing practices that are "transforming how humans relate to each other and nonhuman 'others' and the material environment" (Murris, 2020, p. 44). Rather than being "looked down on" when teachers stand in full stature, children are met at equal standing when teachers change the conditions of bodily intra-actions. Through this posthuman thinking, longstanding perceptions of the identities of children and childhood that are embedded in ideologies of less-than or smaller based on age or stature, are problematised. Disrupting these footholds in thinking metaphorically shakes the ground, exposing new earth for students' identities to be viewed afresh. This "new" vista enables perceptions of children and childhood that are generative, and expansive and pay homage to their innate wisdom by not reducing and limiting their contribution to learning in the classroom context.

Caitlyn extended the agency given to her students, to nonhuman nature, by taking her students outside during lessons and bringing attention to nonhuman others. This was another childhoodnature practice. Caitlyn encouraged interaction with nonhuman nature and gave it emphasis through her own embodied appreciation and respectful relationship. During classes outside, the richness of the agency of nonhuman nature intra-acted with human nature in acts of childhoodnature (Russell & Fawcett, 2020; Stevenson et al., 2020; Waite & Quay, 2020). The affective atmosphere of nature was experienced when Caitlyn took the students outside where they intra-acted with ants, bees, frogs, and the magic and wonder of the material-discursive practices of nonhuman nature. Positioning learning "in place", "on Country", with nonhuman nature, as a community of practice where "place" offers the opportunity to "meld person and world", is an act of childhoodnature (Tooth & Renshaw, 2020). Through such acts as taking students outside to learn, where affect and materiality are used as a mechanism to reconsider human/nonhuman nature relations, anthropocentric dominance is disrupted (Tooth & Renshaw, 2020; Waite & Quay, 2020). Caitlyn understood childhoodnature through its affective atmosphere: one of calmness, less busyness and a sense of settlement and humbleness. The affective atmosphere of nonhuman nature intra-acts with and envelops the human bodies of nature in an entanglement of childhoodnature.

Pedagogies where children engaged *in* outdoor learning provided further evidence of childhoodnature where children were given permission to respond – enact their response-ability – to the material-discursive practices of nature. Outdoor learning was experienced in the lesson participations where learning was undertaken outside – despite the lack of opportunities to engage in wild nature (Blenkinsop et al., 2020; Charles & Louv, 2020; Dyment & Green, 2020; Vladimirova & Rautio, 2020). To enable this, teachers have to contend with and push back against, the material-discursive forces that are commonly experienced by teachers such as a lack of resources, a crowded curriculum, changing educational aims, safety issues, lacking skills, confidence, pedagogical knowledge, and added paperwork (Barfod, 2018; McFarland &

Laird, 2020). In addition, experiences *in* outdoor nature enabled students to engage in more complex and layered intra-actions of childhoodnature through physical proximity and contact with nonhuman others such as bees, frogs, sun, grass, and bark.

The notion of nature *as* teacher (Blom, 2020), co-teacher (Blenkinsop et al., 2020) and diffractive teacher (Murris, 2020), as an act of childhoodnature, was apparent in many of the pedagogical practices. Caitlyn enacted this through listening to and authentically engaging with nonhuman nature. Nonhuman nature was a fundamental part of the classroom and opportunities to learn from nature *as* teacher were consistently enacted, where "learning is understood in terms of different matter – human and nonhuman – making themselves intelligible to each other" (Somerville, 2020, p. 5). Caitlyn equally listened to and respected the agency of nonhuman nature by bringing its voice into attention for her students. Along with her students, she saw awe and magic in nonhuman nature's voice and appreciated the lessons that it naturally offered.

Affective atmospheres were created, through the boundary-making practices or agential cuts, enacted by Caitlyn through their pedagogical choices. It was evident that Caitlyn was highly attuned to the affective atmospheres of her classroom and worked to ensure a steady and consistent affective atmosphere was maintained. The response-ability of teachers in enacting ethical and just decision-making practices acknowledges the nonhuman, material-discursive forces that also intra-act in the classroom milieu and create affective atmospheres (Verlie & Blom, 2021). Affective atmospheres then become an appropriate tool for understanding and qualitatively assessing the material-discursive practices that are being enacted along with the impacts of childhoodnature on the collective student experience.

Methodology as learning: insights from doing, and making

As previously articulated in the Fourth Turn, transqualitative inquiry is an act of reconciliation to fulfil the void left by qualitative and post-qualitative methodologies. In wanting to acknowledge qualitative methods which have been founded on scholarly rigour through the academy and enact them in new and different way, it was found to be impossible to propose an authentic post-qualitative inquiry as put forward by St Pierre (2011).

St Pierre's (2018) post-qualitative methodology refutes structure, order, justification and representation, all of which qualitative research including the conventions of a PhD study, which is where the findings of this research have grown from. In addition, the PhD research informing the findings in this publication was founded on the posthuman thinking of childhoodnature, material-discursive practices and affective atmospheres, which do not align wholly with poststructuralist theories (St Pierre, 2018; Young et al., 2021).

Transqualitative inquiry was here seen to transcend the three traditional research paradigms of quantitative, qualitative, and post-qualitative. This transcendence was a natural progression as the PhD research that informed this book, drawn from the posthuman, is non-interpretive, creative, and yet demonstrates the use of qualitative methods through a posthuman lens. Methodologically, it did not fit the existing paradigms. Just as St Pierre (2011) struggled to fit her own PhD research into qualitative inquiry, thus calling on post-qualitative research, I found a similar rhetoric to justify the need for transqualitative inquiry.

Transqualitative inquiry enabled the exploration of Caitlyn's pedagogical practices across different disciplines. Although the research is focused on environmental education, the teachers were engaged in any discipline of their choice during the lesson participations, which, akin to the transdisciplinary nature of environmental education, provided a fitting space for transqualitative inquiry. Caitlyn engaged in lessons on science and technology, sustainability, and literacy, as there was no restriction placed on what the teachers could teach during the lesson participations. Environmental education is often approached as transdisciplinary and this is an important and necessary "start" (Brown et al., 2020, p. 12). Moreover, Barad (2011, p. 4) argues that,

> The world [is] too complicated for any one set of disciplinary considerations to do justice to the complexity of the issues... [and] a continued insulation of different (inter)disciplinary practices from one another risks missing some crucially entangled epistemological, ontological, and ethical issues.

This may be the case as many studies in the field of science or maths are relegated to quantitative methodologies, and many studies in the arts and humanities are directed to qualitative methodologies; mixed methodologies enable some methodological flexibility; however, they are limited through the human-centric and non-creative approaches that do not allow for challenging theoretical positions. This is perhaps what is needed to expose some of the "entangled epistemological, ontological and ethical issues" (Barad, 2011, p. 4).

Transqualitative inquiry is transitional. Originally, when locating this research into a research paradigm I argued for a methodology that was situated in a liminal space between qualitative and post-qualitative research. At the time I also considered this as a "new" kind of mixed methods approach to research, where the mix was qualitative/post-qualitative. However, there was a theoretical mismatch and these methodologies did not align with the posthumanist positioning of my research along with the creative methods I was employing. In this sense, this research has been transitional. Through the theoretical position of the posthuman collective of childhoodnature, material-discursive practices and affective atmospheres and the methodological approach of diffractive ethnography, this research required a departure, a transition, from existing methodologies. The data created from this study demonstrate a movement away from traditional human-centred thinking by bringing attention and awareness to nonhuman others.

Transqualitative research is transformative. Traditionally, ethnographic studies have focused on the human participants. Indeed, that is the direct translation of ethnography, the study of cultures and as such, the very social structures that humans have created and are a part of. When teamed with diffraction, ethnography is transformed from these deeply embedded human-centred tendencies to include the nonhuman. The human and the nonhuman form the culture *together*. This transformation of methodological approach is in itself a diffraction, as the research created data that highlighted affective atmospheres of vibrant nature. The data identified material-discursive practices that are constantly making themselves known through the teacher's practices. The data highlighted childhoodnature through Caitlyn's pedagogies such as children learning outside with/in/as nature and the refutation of misopedy. These theoretical highlights have transformed the data from being human-centric and brought attention to the nonhuman.

Agential cuts: applying enabling constraints

The data from this book came out of a PhD research project. As such, there are some general applicable limitations such as the four-year full-time equivalent length of the study period, usually without funding. As a primary foray into research, there are learning hurdles to overcome such as the best way to conduct research while discovering the processes of *how* to conduct research at every step while adhering to the required format and traditions of PhD inquiry. In addition, working in educational contexts provides challenges that can also act as limitations such as the time and space availability, interest and subsequent uptake by and of teachers.

In this book, I have shared the data from one of the teachers, Caitlyn, who participated in the study. The perceived limitations of the sample size are considered "enabling constraints" (Manning & Massumi, 2014), which are forced restrictions or "limitative constraints" to an enabler: the limited sample size provided the opportunities for the depth and richness of data generation that has occurred, which would not have otherwise been possible and has positively aligned with an ethnographic methodology (Manning & Massumi, 2014). This "enabling constraint" also had the purpose of bringing attention to the application of the research methodology and methodological approach in practice. Moreover, given the methodological contribution and approach, the data is considered relevant and a vital contribution to environmental education research and education research more broadly.

I also acknowledge that by consenting to partake in the research, Caitlyn may have changed her practices to try and "fit-in" with the "environmental education" focus of my research. That is, Caitlyn may have purposefully planned environmental education lessons in an attempt to show off best practices for the research. This limitation was not mitigated, however, in respect of Caitlyn's professionality and autonomy.

As touched on above, this research was conducted at an unprecedented juncture of the planet: the intra-action of environmental and human devastation associated with the black summer of fires in 2019/2020 (Verlie & Blom, 2021), and the duration of the covid pandemic which began in March 2020 and continues. Further, what has been labelled "the most severe natural disaster in Australia's colonised history" with the East Coast of Australia flooding, has severely impacted the region of this study. Despite these challenges, the research continued, and I am eternally appreciative to Caitlyn for her commitment, flexibility, enthusiasm, and interest in our work.

This research took place in a small rural region in Northern New South Wales, Australia. The region privileges a dominantly Western/minority world perspective with close to 90% of the population speaking only English at home and small percentages of European, Southeast Asian, and Indian populations (.id consulting Pty. Ltd. & Lismore City Council, 2022). Australian First Nations peoples constitute approximately 5% of the population (.id consulting Pty. Ltd. & Lismore City Council, 2022). As was detailed in the Fifth Turn, Caitlyn identifies as an Aboriginal Australian. It is likely that Caitlyn's responses are informed by her cultural background, and this is a prominent feature of ethnographic research. While a limitation, the small sample size affords opportunities for richness and authenticity, as does the ethnographic tradition. As described by Hammersley (2006), central to ethnography is that it is a form of "social and educational research that emphasises the importance of studying firsthand what people do and say in particular contexts" (p. 4). Moreover, ethnography requires lengthy time in the field with each participants' data, which generally includes observation (such as lesson participations) and open-ended interviews (such as video-stimulated recall conversations) to understand "people's perspectives" (Hammersley, 2006, p. 4).

Orientation: future

This book has explored teachers' perceptions of nature and how they inform pedagogy through the posthumanist positioning of the Returning Learning theoretical framework: the entanglement of affective atmospheres, material-discursive practices and childhoodnature. This study provided examples of how environmental education can be reconceptualised to extend beyond exploring different human-perceived mechanisms and approaches to fix, better or solve the global environmental problems that are currently on the planet. Indeed, it is now a time to stop, pause and "bring together" all the learning and knowledge available and apply this to pedagogical approaches. Teachers *are* already practising acts of environmental education, however, the way it has been taught and educated thus far, focuses on a particular way of doing that is siloed into individual subject matter that is fixated on content. Through adopting and applying the theories of childhoodnature, material-discursive practices and affective atmospheres in the Returning Learning framework, and using a diffractive ethnographic trans-qualitative methodology, this book explored and considered how environmental education underpins *every-thing* that teachers do; whether explicitly, implicitly, hidden, or exposed, enactments of alignment to environmental education tenets form part of teachers' every movement. Through conceptualising environmental education from how it is currently, commonly perceived as being a discrete, siloed subject solely responsible for the transfer of content and a regime of practices, into being something that informs teachers' everyday classroom practices and pedagogies – there is much transformation and impact to occur.

To enact pedagogies of environmental education that consider the past, for now, and into the future, the following recommendations generated from this research provide future orientations: (i) for application by teacher practitioners in practice and (ii) for research priorities in environmental education.

Putting the Returning Learning theoretical framework into practice

The Returning Learning theoretical framework is grounded in the theories of material-discursive practices (Barad, 2007), affective atmospheres (Ash & Anderson, 2015) and childhoodnature (Cutter-Mackenzie-Knowles et al., 2020). As has been demonstrated through this book, these theoretical perspectives are highly relevant for teachers to understand and apply in practice. Most pertinent is the need to deconstruct human exceptionalism to consider fracturing the binary of human and nonhuman and to know the human body as nature. The narrative of humans being separate from nonhuman nature still dominates educational spaces and fails to enact a connection to the human body as nature (Blom, 2020). Moreover, through connection to the human body, humans can attune their sensitivities to the affective atmospheres of the classroom, a concept that is heavily theorised through the fields of classroom and school climate (Tomlinson, 2014; Walberg & Anderson, 1968). However, these concepts are problematised due to their anthropocentric and teacher-centric focus (Mayes et al., 2020; Verlie & Blom, 2021) that misses the opportunity to attune to the nonhuman and fails to consider and acknowledge the effect of the material-discursive forces that are influencing and impacting classroom atmospheres and as such, students' and teachers' movements and decision-making practices. This book has argued that affective atmospheres are a powerful mechanism to understand classroom milieus and enable teachers with greater response-ability to their students' bodies as nature and nonhuman nature. Through practising attunement to affective atmospheres, Caitlyn is demonstrated to acknowledge the individuals' needs while moving to ensure a learning space that meets the needs of all students while embracing nonhuman nature as a part of the classroom space. Moreover, through adopting attunement to affective atmospheres and awareness of material-discursive forces, Caitlyn enacted movements that resist misopedy (Murris, 2020) and hold an embodied awareness of the importance of listening to and being in nonhuman nature (Blenkinsop et al., 2020), that is, childhoodnature. Future research that explores, and teacher professional learning that develops, teachers' sensitivities and awareness of material-discursive forces and affective atmospheres in ways to encourage and enhance childhoodnature in classrooms, is urgently needed to empower and inform teachers' decision-making practices in environmental education.

Being responsive not reflective practitioners

There is a body of literature that describes the importance of teachers being culturally and linguistically responsive. However, this book has argued for a "new-kind" of responsivity – one that is attuned to the affective and material-discursive. As demonstrated in this research, teachers like Caitlyn are constantly responding, authentically and in unprepared or practised ways, to their immediate surroundings. Caitlyn was also found to adopt pedagogies that enabled her to respond in ways that fractured material-discursive patterning and forged radically different ways of being a teacher. To do this, Caitlyn pushed back against social norms and configurations of their "teacherly selves", their identity as a teacher through reimagining what it means to be a teacher. This responsivity, as conceptualised through Barad's (2007) quantum theorising, troubled the fixation and obsession that education currently has with reflection. Posthuman thinkers such as Haraway and Barad have problematised the focus on reflection primarily for its affinity to highlight "more of the same" rather than enacting diffractive movements that actually afford *new* knowledge generation. In the context of higher education and reflective practice with medical students, Mitchell (2017) adds that "diffraction takes reflective practices beyond the private realm and the binaries of student/educator, theory/practice, experience/writing. New possibilities are invoked towards responding to social injustices ... thereby empowering and motivating students to become change makers" (p. 180).

Moreover, in the process of "looking back", teachers focus on what they have done before to recreate "good" pedagogies and alleviate the "bad". However, this objective position denies the quantum changes that are constantly occurring. Moreover, in looking back, teachers are blind to what is happening in front of them. Attunement, through a deep connection to the body as nature, enacts an ability to respond to the body's call. I assert that future research that delves more deeply into how diffractive and responsive practices can be implemented by teachers instead of reflective practices, would be timely and beneficial.

Conclusion

Throughout this book, I have explored teachers' perceptions of nature and how they inform pedagogy. This foray into environmental education research has demonstrated the richness and depth of one teacher's perception and pedagogy, which has provided a crucial contribution to knowledge in these uncertain times.

Teachers' perceptions of nature are a grossly under-researched phenomenon. There has been a dearth of research that specifically seeks to find out *how* teachers perceive nature and *how* this perception informs their pedagogy. Through this study, it was found that Caitlyn perceived nature as something external and separate from her own body. However, it is also noted that there was a general disconnect between teachers' perception of nature in the verbal/conversational and practical contexts as often an anthropocentric position was perpetuated that refuted posthuman thinking of the human body *as* nature.

The movement into this posthuman way of thinking, and as such, way of educating, offers new ways of conceptualising environmental education that attend more respectfully and relationally to the current global, environmental disturbances and emergencies that human and nonhuman nature are enduring. Examples of catastrophic global changes have been frequent in recent times, and in particular, these embodied experiences have been relentless during the period of this research for myself and Caitlyn, including the 2019 black summer of fires, the covid pandemic and the 1-in-3,500-year flooding incident. This research provides a timely and significant contribution to address these unprecedented events and those yet to come: the uncertainties, unpredictabilities, and seeming impossibilities of the planet's future scape.

In this book, I have argued that there is no doubt that the current time signifies a critical point in the future of the planet, as both human and nonhuman nature are equally implicated by the choices that humans make at this crucial juncture. Posthuman thinking, as described in this thesis through the concepts of material-discursive practice, affective atmospheres and childhoodnature, provides a radically *new* way of approaching environmental education in shifting the anthropocentric obsession to an ethico-onto-epistemology that embodies perceptions of human bodies as nature as an act of social justice for a future to come.

To conclude, I return to the voice of the teacher with two vignettes from Caitlyn:

> *We do it like*
> *"don't play with sticks,*
> *don't climb the trees,*
> *don't do this"...*
> *Where is the risk taking and*
> *that connection to being outdoors*
> *and in nature*
> *and just being kids?*[1]

> *I find I'm at my best,*
> *or when I reflect and I feel like I'm their best teacher*
> *is when I'm in an organic mode...*
> *Yeah you've got to stick to your structure, your curriculum, your*
>   *criteria...*
> *you've got to tick those boxes,*
> *but I find when it's flowing,*
> *or then tomorrow it's like yeah let's do this,*
> *scrap what I planned...let's take it in a different direction.*
> *That's when I feel like they're having the best time,*
> *they're learning more,*
> *and I just feel like a much better teacher.*[2]

Notes

1 Caitlyn, video-stimulated recall conversation, 11th June 2020.
2 Caitlyn, video-stimulated recall conversation, 19th February 2020.

References

.id consulting Pty. Ltd., & Lismore City Council. (2022). *Lismore City community profile*. https://profile.id.com.au/lismore/language#o= place&s=northern%20rivers%20nsw

Ash, J., & Anderson, B. (2015). Atmospheric methods. In P. Vannini (Ed.), *Non-representational methodologies* (pp. 44–61). Routledge.

Barad, K. (2007). *Meeting the universe halfway: Quantum physics and the entanglement of matter and meaning*. Duke University Press.

Barad, K. (2011). Erasers and erasures: Pinch's unfortunate "Uncertainty Principle". *Social Studies of Science, 41*(3), 443–454.

Barfod, K. S. (2018). Maintaining mastery but feeling professionally isolated: Experienced Teachers' perceptions of teaching outside the classroom. *Journal of Adventure Education and Outdoor Learning, 18*(3), 201–213.

Bennett, J. (2010). *Vibrant matter: A political ecology of things*. Duke University Press.

Blenkinsop, S., & Piersol, L. (2013). Listening to the literal: Orientations towards how nature communicates. *Phenomenology & Practice, 7*(2), 41–60.

Blenkinsop, S., Jickling, B., Morse, M., & Jensen, A. (2020). Wild pedagogies: Six touchstones for childhoodnature theory and practice. In A. Cutter-Mackenzie, K. Malone, & E. Barratt Hacking (Eds.), *Research handbook on childhoodnature: Assemblages of childhood and nature research* (pp. 1–18). Springer International Publishing.

Blom, S. M. (2016). *Parent(ing) childhoodnature: Perceptions with/as nature*. Southern Cross University.

Blom, S. M. (2020). Conceptualizing parent(ing) childhoodnature through significant life experience. In A. Cutter-Mackenzie, K. Malone, & E. Barratt Hacking (Eds.), *Research handbook on childhoodnature: Assemblages of childhood and nature research* (pp. 1–26). Springer International Publishing.

Bonnett, M. (2002). Education for sustainability as a frame of mind. *Environmental Education Research, 8*(1), 9–20. https://doi.org/10.1080/13504620120109619.

Brown, S. L., Siegel, L., & Blom, S. M. (2020). Entanglements of matter and meaning: The importance of the philosophy of Karen Barad for environmental education. *Australian Journal of Environmental Education, 36*(3), 1–15.

Carson, R. (1965). *Sense of wonder*. Harper & Row.

Charles, C., & Louv, R. (2020). Wild hope: The transformative power of children engaging with nature. *Research handbook on childhoodnature: Assemblages of childhood and nature research* (pp. 1–21). Springer.

Chawla, L. (1998). Significant life experiences revisited: A review of research. *Journal of Environmental Education, 29*(3), 11.

Cutter-Mackenzie, A., & Smith, R. (2003). Ecological literacy: The "Missing Paradigm" in environmental education (Part one). *Environmental Education Research, 9*(4), 497–524.

Cutter-Mackenzie-Knowles, A., Malone, K., & Barratt Hacking, E. (2020). *Research handbook on childhoodnature: Assemblages of childhood and nature research*. Springer International Publishing.

Desjean-Perrotta, B., Moseley, C., & Cantu, L. E. (2008). Preservice Teachers' perceptions of the environment: Does ethnicity or dominant residential experience matter? *The Journal of Environmental Education, 39*(2), 21–32.

Dyment, J., & Green, M. (2020). Everyday, local, nearby, healthy childhoodnature settings as sites for promoting children's health and well-being. In A. Cutter-Mackenzie-Knowles, K. Malone, & E. Barratt Hacking (Eds.), *Research handbook on childhoodnature: Assemblages of childhood and nature research* (pp. 1–26). Springer International Publishing.

Fraser, J., Gupta, R., & Krasny, M. E. (2015). Practitioners' perspectives on the purpose of environmental education. *Environmental Education Research, 21*(5), 777–800.

García-González, E., Jiménez-Fontana, R., & Azcárate, P. (2020). Education for sustainability and the sustainable development goals: Pre-service teachers' perceptions and knowledge. *Sustainability, 12*(18), 7741.

Greenwood, D. A. (2020). Rachel Carson's childhood ecological aesthetic and the origin of the sense of wonder. In A. Cutter-Mackenzie-Knowles,

K. Malone, & E. Barratt-Hacking (Eds.), *Research handbook on child-hoodnature: Assemblages of childhood and nature research* (pp. 1639–1656). Springer International Publishing.

Hammersley, M. (2006). Ethnography: Problems and prospects. *Ethnography and Education*, *1*(1), 3–14.

Kalvaitis, D., & Monhardt, R. (2015). Children voice biophilia: The phenomenology of being in love with nature. *The Journal of Sustainability Education*, *9*. https://www.susted.com/wordpress/content/children-voice-biophilia-the-phenomenology-of-being-in-love-with-nature_2015_03/

Lingard, B., Hayes, D., & Mills, M. (2003). Teachers and productive pedagogies: Contextualising, conceptualising, utilising. *Pedagogy, Culture & Society*, *11*(3), 399–424. https://doi.org/10.1080/14681360300200181.

Loughland, T., Reid, A., & Petocz, P. (2002). Young people's conceptions of environment: A phenomenographic analysis. *Environmental Education Research*, *8*(2). https://doi.org/10.1080/13504620220128248.

Louv, R. (2006). *Last child in the woods: Saving our children from nature-deficit disorder*. Atlantic Books.

Malone, K. (2007). The bubble-wrap generation: Children growing up in walled gardens. *Environmental Education Research*, *13*(4), 513–527. https://doi.org/10.1080/13504620701581612.

Manning, E., & Massumi, B. (2014). *Thought in the act: Passages in the ecology of experience*. University of Minnesota Press.

Mayes, E., Wolfe, M. J., & Higham, L. (2020). Re/imagining school climate: Towards processual accounts of affective ecologies of schooling. *Emotion, Space and Society*, *36*, 100703.

McFarland, L., & Laird, S. G. (2020). "She's only two": Parents and educators as gatekeepers of children's opportunities for nature-based risky play. *Research handbook on childhoodnature: Assemblages of childhood and nature research* (pp. 1075–1098). Springer International Publishing.

Mitchell, V. A. (2017). Diffracting reflection: A move beyond reflective practice. *Education as Change*, *21*, 165–186.

Murris, K. (2020). Posthuman child and the diffractive teacher: Decolonizing the nature/culture binary. In A. Cutter-Mackenzie, K. Malone, & E. Barratt Hacking (Eds.), *Research handbook on childhoodnature: Assemblages of childhood and nature research* (pp. 1–25). Springer International Publishing.

O'Riordan, T. (1977). Environmental ideologies. *Environment and Planning A, 9*(1), 3–14.

Payne, P. (1998). Children's conceptions of nature. *Australian Journal of Environmental Education, 14,* 7.

Quinn, F., Castéra, J., & Clément, P. (2016). Teachers' conceptions of the environment: anthropocentrism, non-anthropocentrism, anthropomorphism and the place of nature. *Environmental Education Research, 22*(6), 893–917. https://doi.org/10.1080/13504622.2015.1076767.

Rousell, D., & Cutter-Mackenzie-Knowles, A. (2020). Uncommon worlds: Toward an ecological aesthetics of childhood in the Anthropocene. In A. Cutter-Mackenzie-Knowles, K. Malone, & E. Barratt Hacking (Eds.), *Research handbook on childhoodnature: Assemblages of childhood and nature research* (pp. 1–23). Springer International Publishing.

Russell, J., & Fawcett, L. (2020). *Childhood animalness: Relationality, vulnerabilities, and conviviality.* In A. Cutter-Mackenzie-Knowles, K. Malone, & E. Barratt Hacking (Eds.), *Research handbook on childhoodnature: Assemblages of childhood and nature research* (pp. 1339–1354). Springer International Publishing. https://doi.org/10.1007/978-3-319-67286-1_64

Shepardson, D. P. (2005). Student ideas: What is an environment? *The Journal of Environmental Education, 36*(4), 49.

Simmons, D. (1993). Facilitating teachers' use of natural areas: Perceptions of environmental education opportunities. *The Journal of Environmental Education, 24*(3), 8–16.

Sobel, D. (1996). *Beyond ecophobia. Reclaiming the heart in nature education.* Orion Society.

Sobel, D. (2008). *Childhood and nature: Design principles for educators.* Sternhouse Publishers.

Somerville, M. (2020). Posthuman theory and practice in early years learning. In A. Cutter-Mackenzie, K. Malone, & E. Barratt Hacking (Eds.), *Research handbook on childhoodnature: Assemblages of childhood and nature research* (pp. 1–25). Springer International Publishing.

St. Pierre, E. A. (2011). Post qualitative research: The critique and the coming after. *The Sage handbook of qualitative research* (4th ed., pp. 611–626). Sage.

St Pierre, E. A. (2018). Writing post-qualitative inquiry. *Qualitative Inquiry, 24*(9), 603–608. https://doi.org/10.1177/1077800417734567.

Stevenson, R. B., Mannion, G., & Evans, N. (2020). Childhoodnature pedagogies and place: An overview and analysis. In A. Cutter-Mackenzie-Knowles, K. Malone, & E. Barratt Hacking (Eds.), *Research handbook on childhoodnature: Assemblages of childhood and nature research* (pp. 1401–1421). Springer International Publishing. https://doi.org/10.1007/978-3-319-67286-1_64

Taylor, C. A. (2013). Objects, bodies and space: Gender and embodied practices of mattering in the classroom. *Gender and Education, 25*(6), 688–703. https://doi.org/10.1080/09540253.2013.834864.

Thi To Khuyen, N. G. U. Y. E. N., Van Bien, N. G. U. Y. E. N., Lin, P.-L., Lin, J., & Chang, C.-Y. (2020). Measuring Teachers' perceptions to sustain stem education development. *Sustainability, 12*(4), 1531.

Tomlinson, C. A. (2014). *The differentiated classroom: Responding to the needs of all learners.* ASCD.

Tooth, R., & Renshaw, P. (2020). Children becoming emotionally attuned to "nature" through diverse place-responsive pedagogies. In A. Cutter-Mackenzie, K. Malone, & E. Barratt Hacking (Eds.), *Research handbook on childhoodnature: Assemblages of childhood and nature research* (pp. 1–22). Springer International Publishing.

Torquati, J., Cutler, K., Gilkerson, D., & Sarver, S. (2013). Early childhood Educators' perceptions of nature, science, and environmental education. *Early Education and Development, 24*(5), 721–743. https://doi.org/10.1080/10409289.2012.725383.

Verlie, B., & Blom, S. M. (2021). Education in a changing climate: Reconceptualising school and classroom climate through the fiery atmos-fears of Australia's black summer. *Children's Geographies*, 1–15. https://doi.org/10.1080/14733285.2021.1948504.

Vladimirova, A., & Rautio, P. (2020). Unplanning research with a curious practice methodology: Emergence of childrenforest in the context of Finland. *Research handbook on childhoodnature: Assemblages of childhood and nature research* (pp. 1–26). Springer International Publishing.

Waite, S., & Quay, J. (2020). In place(s): Dwelling on culture, materiality, and affect. In A. Cutter-Mackenzie-Knowles, K. Malone, & E. Barratt Hacking (Eds.), *Research Handbook on childhoodnature: Assemblages of childhood and nature research* (pp. 179–198). Springer International Publishing. https://doi.org/10.1007/978-3-319-67286-1_64

Walberg, H. J., & Anderson, G. J. (1968). Classroom climate and individual learning. *Journal of Educational Psychology, 59*(6p1), 414.

Weston, A. (2004). What if teaching went wild? *Canadian Journal of Environmental Education, 9*(1), 31–46.

Young, T., Crinall, S., & Malone, K. (2021). Disruptions of post-qualitative education research: Tensions and openings. *Qualitative Inquiry, 28*(3–4), 312–321.

INDEX

Note: Page references in *italics* denote figures and with "n" endnotes.